VOLUME ONE HUNDRED AND TWENTY FIVE

ADVANCES IN
COMPUTERS
Data Prefetching Techniques
in Computer Systems

VOLUME ONE HUNDRED AND TWENTY FIVE

ADVANCES IN
COMPUTERS
Data Prefetching Techniques
in Computer Systems

Edited by

PEJMAN LOTFI-KAMRAN
School of Computer Science,
Institute for Research in Fundamental Sciences (IPM),
Tehran, Iran

HAMID SARBAZI-AZAD
Sharif University of Technology, and
Institute for Research in Fundamental Sciences (IPM),
Tehran, Iran

ACADEMIC PRESS
An imprint of Elsevier

ELSEVIER

Academic Press is an imprint of Elsevier
50 Hampshire Street, 5th Floor, Cambridge, MA 02139, United States
525 B Street, Suite 1650, San Diego, CA 92101, United States
The Boulevard, Langford Lane, Kidlington, Oxford OX5 1GB, United Kingdom
125 London Wall, London, EC2Y 5AS, United Kingdom

First edition 2022

ISBN: 978-0-323-85119-0
ISSN: 0065-2458

For information on all Academic Press publications
visit our website at https://www.elsevier.com/books-and-journals

Publisher: Zoe Kruze
Developmental Editor: Cindy Angelita Gardose
Production Project Manager: James Selvam
Cover Designer: Greg Harris

Typeset by STRAIVE, India

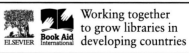

Working together
to grow libraries in
developing countries

www.elsevier.com • www.bookaid.org

Contents

Contributors

Ali Ansari
Sharif University of Technology, Tehran, Iran

Fatemeh Golshan
Sharif University of Technology, Tehran, Iran

Pejman Lotfi-Kamran
School of Computer Science, Institute for Research in Fundamental Sciences (IPM), Tehran, Iran

Hamid Sarbazi-Azad
Sharif University of Technology, and Institute for Research in Fundamental Sciences (IPM), Tehran, Iran

Mehran Shakerinava
Sharif University of Technology, Tehran, Iran

Preface

Advances in Computers, the oldest series to the chronicle of the rapid evolution of computing, annually publishes several volumes, each comprising typically four to eight chapters, describing new findings and developments in the theory and applications of computing.

This 125th volume is a thematic one entitled "Data Prefetching Techniques in Computer Systems" inspired by recent work in data prefetching techniques proposed for advanced processors and used in computer systems. This volume comprises six chapters.

While processor performance has been rapidly improving during the past decades, the main goal of designers was to increase the capacity of the main memory rather than its speed. Therefore, there emerged a considerable gap, called memory wall, between the speed of the processor and main memory. To tackle the memory wall problem, a high-speed low-capacity buffer called cache memory is used between the processor and the main memory to remedy their performance gap. When a processor requires a piece of data, if it is in the cache (called a hit), it can be delivered to the processor very quickly. Otherwise, which is called a miss, the data has to be accessed in the main memory. In such cases, cache memory is not useful and even increases the access latency as we first need to search the cache and then access the main memory. As the main memory latency is considerable for a processor, lowering the requests served by the memory is desirable. Data prefetching is a way to improve the effectiveness of caches and reduces the performance gap between the processor and the main memory by lowering the stall cycles seen by the processor when serving main memory requests. A data prefetcher predicts that a piece of data is going to be used by the processor. In case the predicted data is not in the cache, the data prefetcher brings it into the cache in advance.

The main task of a data prefetcher is to predict future memory accesses. Fortunately, the memory accesses of a processor usually are regular and follow patterns. By learning access patterns, a data prefetcher can make correct predictions. Based on different access patterns that a data prefetcher can learn, different types of data prefetchers may be designed and used. This book introduces several important types of data prefetchers and discusses how they perform data prefetching, why they work as intended, when they are effective, and when they are not.

In this book, we first discuss the basic definitions and fundamental concepts of data prefetching and then study classic hardware data prefetchers, as well as recent advanced data prefetchers. We describe the operations of every data prefetcher in detail and shed light on its design trade-offs.

The six chapters of the book are organized as follows: Chapter 1, an introduction to data prefetching is given. Different prefetchers are defined and many recent prefetchers are categorized as well. Chapter 2 discusses spatial prefetching techniques and introduces SMS and VLDP techniques in detail. Chapter 3 discusses temporal data prefetching and explains STMS and ISB. Techniques that cannot be classified under only spatial and temporal data prefetchers are considered in Chapter 4. Chapter 5 introduces some recently proposed data prefetchers. Finally, Chapter 6 evaluates the effectiveness of different prefetchers under various working conditions.

We hope that the readers will find this volume interesting and useful for teaching, research, and designing data prefetchers. We welcome any feedback and suggestions related to the book.

PEJMAN LOTFI-KAMRAN
School of Computer Science
Institute for Research in Fundamental Sciences (IPM)
Tehran, Iran

HAMID SARBAZI-AZAD
Sharif University of Technology,
and
Institute for Research in Fundamental Sciences (IPM),
Tehran, Iran

Table of abbreviations

AMT access map table
BOP best-offset prefetcher
CPU central processing unit
DHB delta history buffer
DPC data prefetching championship
DPT delta prediction table
DRAM dynamic random access memory
DSS decision support system
EIT enhanced index table
GPU graphics processing unit
IPC instructions per cycle
ISB irregular stream buffer
L1 D level 1 data
LLC last-level cache
LRU least recently used
MLOP multi-lookahead offset prefetcher
NVM non-volatile memory
OLTP online transaction processing
OoO out of order
OPT offset prediction table
OS operating system
PAS physical address space
PC program counter
PHT pattern history table
PSAM physical-to-structural address mapping
PST patterns sequence table
RMD runahead metadata
RMOB region miss order buffer
ROB reorder buffer
RRT recent requests table
SAS structural address space
SMS spatial memory streaming
SP sandbox prefetcher
SPAM structural-to-physical address mapping
SPU sandbox prefetch unit
STeMS spatio-temporal memory streaming
STMS sampled temporal memory streaming
TLB translation lookaside buffer
TPC transaction processing performance council
VLDP variable length delta prefetcher

CHAPTER ONE

Introduction to data prefetching

Pejman Lotfi-Kamran[a] and Hamid Sarbazi-Azad[b]

[a]School of Computer Science, Institute for Research in Fundamental Sciences (IPM), Tehran, Iran
[b]Sharif University of Technology, and Institute for Research in Fundamental Sciences (IPM), Tehran, Iran

Contents

Abstract

For many years, computer designers benefitted from Moore's law to significantly improve the speed of processors. Unlike processors, with the main memory, the focus of improvement has been capacity and not access time. Hence, there is a large gap between the speed of processors and main memory. To hide the gap, a hierarchy of caches has been used between a processor and main memory. While caches proved to be quite effective, their effectiveness directly depends on how many times a requested piece of data can be found in the cache. Due to the complexity of data access patterns, on many occasions, the requested piece of data cannot be found in the cache hierarchy, which exposes large delays to processors and significantly degrades their performance. Data prefetching is the science and art of predicting, in advance, what pieces of data a processor needs and bringing them into the cache before the processor requests them. In this chapter, we justify the importance of data prefetching, look at a taxonomy of data prefetching, and briefly discuss some simple ways to do prefetching.

1. Introduction

We are living in an era in which Information Technology (IT) is shaping our society. More than any time in history, our society is dependent on IT for its day-to-day activities. Education, media, science, social networking, etc. are all affected by IT [1]. The steady growth in processor performance is one of the driving forces behind the success and widespread adoption of IT.

Historically, the processor performance had been improved by a factor close to two every 2 years (this phenomenon sometimes mistakenly is referred to as Moore's law [2]). While processor performance was rapidly improving, the goal of designers was to rapidly increase the capacity of the DRAM, which is the building block of main memory. As the speed of DRAM was a second-level optimization for designers, over time, there emerged a considerable gap between the speed of processors and main memory to which many refer as *memory wall*.

Since 2004, the rate at which the processor performance is improving has been reduced [3]. Due to the reduction in the annual improvement in processor performance, the memory wall is not widening as in the last two decades of the 20th century. Nevertheless, the memory wall is considerable and hurts processor performance.

To hide the delay of the main memory, historically, a cache is used between the processor and the main memory. A cache is a small and fast hardware-managed storage build from SRAM, which is much faster and requires more space than DRAM. As cache size is very small, only part of the data can be stored in the cache at any given point in time. When a processor asks for a piece of data, if it is in the cache (called a hit), it can be delivered to the processor very quickly. However, if the data is not in the cache (called a miss), we have to get the data elsewhere (e.g., main memory). In such cases, not only the cache is not useful but also it increases the access latency as we first need to search the cache and then look for the piece of data elsewhere.

To increase the likelihood of finding the data in the cache, contemporary processors benefit from a hierarchy of caches. These processors use multiple levels of caches. The cache that is closest to the processor is the smallest and the fastest. We usually refer to this cache as the L1 cache. As we further away from the processor, the caches become larger and slower. We refer to these caches as L2, L3, etc. The cache at the end of the cache hierarchy, which is closest to the main memory, is usually referred to as the *last level cache* (LLC).

The cache hierarchy is quite effective and serves many of the processor memory requests. Nonetheless, the cache hierarchy cannot handle all of the memory requests. For every request that goes to the main memory, the processor is exposed to the whole delay of accessing memory. Given that the delay of main memory is considerable for a processor, a mechanism that lowers the number of requests served by the memory is quite useful.

Data prefetching is a way to improve the effectiveness of caches and lower the stall cycles due to serving processor requests by the main memory. A data prefetcher predicts that a piece of data is going to be used by the processor. Based on this prediction, in case the predicted data is not in the cache, the data prefetcher brings it into the cache. If the prediction is correct, a cache miss is avoided, and the processor finds the data in the cache when it is needed. Otherwise, a useless piece of data is brought into the cache, and possibly a useful piece of data is evicted from the cache to open room for the predicted data.

The main part of a data prefetcher is how to predict future memory accesses of a processor. Fortunately, the memory accesses of a processor usually are regular and follow patterns (as we see in this book). By learning a pattern and using it for prediction, a data prefetcher can make correct predictions.

There are many patterns that a data prefetcher may learn, and hence, there are many different types of data prefetchers. In this book, we introduce several important types of data prefetchers. We discuss how they perform data prefetching and why they work as intended. We also mention when they are effective and when they are not. In this book, we first discuss the fundamental concepts in data prefetching then study recent, as well as classic, hardware data prefetchers. We describe the operations of every data prefetcher, in detail, and shed light on its design trade-offs.

2. Background

In this section, we briefly review some background on hardware data prefetching.

2.1 Predicting memory references

The first step in data prefetching is predicting future memory accesses. Fortunately, data accesses demonstrate several types of correlations and localities, which lead to the formation of patterns among memory accesses, from which data prefetchers can predict future memory references. These patterns emerge from the layout of programs' data structures in the memory, and the algorithm and the high-level programming constructs that operate on these data structures. In this chapter, we briefly mention three important memory access patterns of applications: (1) stride, (2) temporal, and (3) spatial access patterns.

2.1.1 Stride accesses

Stride access pattern refers to a sequence of memory accesses in which the distance of consecutive accesses is constant, e.g., $\{A, A+k, A+2k, \ldots\}$ with stride k. Such patterns are frequent in programs with dense matrices and frequently come into sight when programs operate on multi-dimensional arrays. Please consider the following piece of code.
Pseudocode 1. A simple program that calculates the sum of an array of bytes.

```
byte A [100]
// some computation with A
...
SUM = 0
for (int i = 0; i < 100; i++)
    SUM += A[i]
```

In this example, all elements of array A are read to be added to the variable SUM, which holds the aggregate sum of array A. As all elements of an array are placed one after another in memory, if $A[0]$ is placed at address $addr$, $A[1]$ is placed at $addr + 1$. Similarly, the rest of the elements are placed at $addr + 2, addr + 3, \ldots, addr + 99$. Due to this particular placement of array A in the memory, and the nature of the algorithm, Pseudocode 1, during execution, generates memory references $addr, addr + 1, \ldots,$ and $addr + 99$, and hence exhibits a simple stride access pattern with a stride of 1. If we replace the "for $(int\ i = 0;\ i < 100;\ i++)$" statement in the code to "for $(int\ i = 0;$ $i < 100;\ i += 2)$," the code references addresses $addr, addr + 2, addr + 4,$ $\ldots, 0\ addr + 98$, and hence exhibits an access pattern with a stride of 2.

While stride accesses are abundant in array and matrix data structures, they are not unique to such data structures. Stride accesses also appear in pointer-based data structures when memory allocators arrange the objects sequentially and in a constant-size manner in the memory [1].

2.1.2 Temporal address correlation

Temporal address correlation [2] refers to a sequence of addresses that favor being accessed together and in the same order. For example, if we observe

{A,B,C,D}, then it is likely for {B,C,D} to follow {A} in the future. Temporal address correlation stems fundamentally from the fact that programs consist of loops, and is observed when data structures such as lists, arrays, and link lists are traversed. When data structures are stable [4], access patterns recur, and the temporal address correlation is manifested [2]. Please consider the following piece of code.

Pseudocode 2. A simple program that calculates the sum of a linked list of bytes.

```
struct ELEMENT_T {byte B; pointer P};
ELEMENT_T * e;
// create and manupulate a link list of ELEMENT_T that e points to
...
ELEMENT_T * p = e;
while (p != NULL)
   SUM += p.B;
   p = p.pointer
```

The functionality of Pseudocode 2 is similar to Pseudocode 1. The major difference is due to replacing the array in Pseudocode 1 with a linked list. As every element of a linked list can be anywhere in the memory, there is usually no manifestation of a stride access pattern when the code is being executed. However, if we traverse the linked list multiple time, e.g., the code makes some modification in the linked list and then attempts to determine the total sum, the sequence of memory accesses due to linked in traversal will be the same as the previous time except for differences due to addition or deletion of certain elements in the linked list. If the change in the linked list is not significant, most of the sequence of memory references are identical across multiple traversals of the linked list, which lead to temporal access pattern manifestation.

2.1.3 Spatial address correlation

Spatial address correlation [3] refers to the phenomenon that similar access patterns occur in different regions of memory. For example, if a program visits locations {A,B,C,D} of page X, it is probable that it visits locations {A,B,C,D} of other pages as well. Spatial correlation transpires because applications use various objects with a regular and fixed layout, and accesses reappear while traversing data structures [3]. To better understand the concept, please consider the following piece of code.

Pseudocode 3. A simple program that calculates the sum of two elements of a linked list.

---

struct ELEMENT_T {byte a; byte A1 [100]; byte b; byte A2 [20]; byte c; pointer p;}
*ELEMENT_T * e;*
// create and manupulate a link list of ELEMENT_T that e points to
...
*ELEMENT_T * p = e;*
while (p != NULL).
   SUM += c;
   SUM += b;
   SUM += a;
   p = p.pointer

The code is similar to Pseudocode 2. The major difference is that each element of the linked list is a large structure consisting of two arrays and three scalar variables in addition to the pointer. The piece of code, however, only cares about the three scalar variables and attempts to calculate the sum of all the scalar variables in the linked list. As the data structure is a linked list and hence all of its elements may be in different locations in the main memory, the code likely does not exhibit stride access patterns. Moreover, unlike the last example, assume that we plan to traverse the linked list once. As such, the code does not exhibit temporal access patterns as well. Nonetheless, while each element of the linked list might be in any location in the main memory, once the code touches variable c, it also touches variables b and a. Moreover, as these variables are located in the main memory with a fixed layout dictated by the ELEMENT_T structure, when the first element is touched, the location of the other two can be inferred. This access pattern is called a spatial access pattern.

2.2 Prefetching lookahead

Prefetchers need to issue timely prefetch requests for the predicted addresses. Preferably, a prefetcher sends prefetch requests well in advance and supplies enough storage for the prefetched blocks to hide the entire latency of memory accesses. An early prefetch request may cause evicting a useful block from the cache, and a late prefetch may decrease the effectiveness of prefetching in that a portion of the long latency of memory access is exposed to the processor.

Prefetching lookahead refers to how far ahead of the demand miss stream the prefetcher can send requests. An aggressive prefetcher may offer a high prefetching lookahead (say, 8) and issue many prefetch requests ahead of the processor demand requests to hide the entire latency of memory accesses; on the other hand, a conservative prefetcher may offer a low prefetching lookahead and send a single prefetch request in advance of the processor's demand to avoid wasting resources (e.g., cache storage and memory bandwidth). Typically, there is a trade-off between the aggressiveness of a prefetching technique and its accuracy: making a prefetcher more aggressive usually leads to covering more data–miss–induced stall cycles but at the cost of fetching more useless data.

Some pieces of prior work propose to dynamically adjust the prefetching lookahead [5–7]. Based on the observation that the optimal prefetching degree is different for various applications and various execution phases of a particular application, as well, these approaches employ heuristics to increase or decrease the prefetching lookahead. For example, SPP [6] monitors the accuracy of issued prefetch requests and reduce the prefetching lookahead if the accuracy becomes smaller than a predefined threshold.

2.3 Location of data prefetcher

Prefetching can be employed to move the data from lower levels of the memory hierarchy to any higher level. Prior work used data prefetchers at all cache levels, from the primary data cache to the shared last-level cache.

The location of a data prefetcher has a profound impact on its overall behavior [8]. A prefetcher in the first-level cache can observe all memory accesses, and hence, can issue highly accurate prefetch requests, but at the cost of imposing large storage overhead for recording the metadata information. In contrast, a prefetcher in the last-level cache observes the access sequences that have been filtered at higher levels of the memory hierarchy, resulting in lower prediction accuracy, but higher storage efficiency.

2.4 Prefetching hazards

A naive deployment of a data prefetcher not only may not improve the system performance but also may significantly harm the performance and energy efficiency [9]. The two well-known major drawbacks of data prefetching are (1) cache pollution and (2) off-chip bandwidth overhead.

2.4.1 Cache pollution

Data prefetching may increase the demand misses by replacing useful cache blocks with useless prefetched data, harming the performance. Cache pollution usually occurs when an aggressive prefetcher exhibits low accuracy and/or when prefetch requests of a core in a many-core processor compete for shared resources with demand accesses of other cores [10].

2.4.2 Bandwidth overhead

In a many-core processor, prefetch requests of a core can delay demand requests of another core because of contending for memory bandwidth [10]. This interference is the major obstacle of using data prefetchers in many-core processors, and the problem gets thornier as the number of cores increases [11,12].

2.4.3 Placing prefetched data

Data prefetchers usually place the prefetched data into one of the following two structures: (1) the cache itself, and (2) an auxiliary buffer next to the cache. In case an auxiliary buffer is used for the prefetched data, demand requests first look for the data in the cache; if the data is not found, the auxiliary buffer is searched before sending a request to the lower levels of the memory hierarchy.

Storing the prefetched data into the cache lowers the latency of accessing data when the prediction is correct. However, when the prediction is incorrect or when the prefetch request is not timely (i.e., too early), having the prefetched data in the cache may result in evicting useful cache blocks.

2.5 Prefetcher types

There are several types of data prefetchers. At a very high level, data prefetchers can be classified into hardware prefetchers and nonhardware prefetchers. A hardware prefetcher is a data prefetching technique that is implemented as a hardware component in a processor. Any other prefetching technique is a nonhardware prefetcher. Fig. 1 shows a classification of data prefetching techniques.

We focus on hardware data prefetching techniques in this book. As shown in Fig. 1, Hardware data prefetchers can be classified into spatial, temporal, and nonspatial-temporal prefetchers. We cover conventional spatial prefetchers in chapter "Spatial prefetching" by Lotfi-Kamran and Sarbazi-Azad and state-of-the-art spatial prefetchers in chapter "State-of-the-art data prefetchers" by Shakerinava et al. Conventional and state-of-

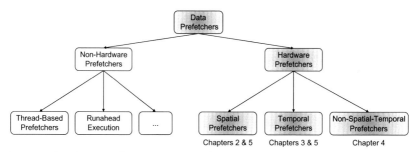

Fig. 1 A classification of various data prefetching techniques (with an emphasis on hardware data prefetchers).

the-art temporal prefetchers are covered in chapters "Temporal prefetching" by Lotfi-Kamran and Sarbazi-Azad; "State-of-the-art data prefetchers" by Shakerinava et al., respectively. Non-spatial-temporal prefetchers are covered in chapter "Beyond spatial or temporal prefetching" by Lotfi-Kamran and Sarbazi-Azad. We evaluate various types of hardware data prefetchers in chapter "Evaluation of data prefetchers" by Shakerinava et al. to empirically assess their strengths and weaknesses.

Fig. 1 also shows a classification of the non-hardware data prefetchers. As this book is about hardware data prefetching, we do not cover non-hardware data prefetching techniques. However, Section 5 of this chapter reviews various non-hardware data prefetching techniques and offers a short explanation for them.

3. A preliminary hardware data prefetcher

To give insight on how a stereotype operates, now we describe a preliminary-yet-prevalent type of stride prefetching. Generally, stride prefetchers are widely used in commercial processors (e.g., IBM Power4 [13], Intel Core [14], AMD Opteron [15], Sun UltraSPARC III [16]) and have been shown quite effective for desktop and engineering applications. Stride prefetchers [4,17–23] detect streams (i.e., the sequence of consecutive addresses) that exhibit stride access patterns and generate prefetch requests by adding the detected stride to the last observed address.

Instruction-Based Stride Prefetcher (IBSP) [17] is a preliminary type of stride prefetching. The prefetcher tracks the stride streams on a per load instruction basis: the prefetcher observes accesses issued by individual load instructions and sends prefetch requests if the accesses manifest a stride pattern. Fig. 2 shows the organization of IBSP's metadata table, named

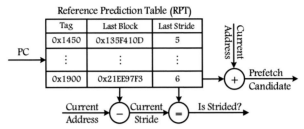

Fig. 2 The organization of Instruction-Based Stride Prefetcher (IBSP).The 'RPT' keeps track of various streams.

Reference Prediction Table (RPT). RPT is a structure tagged and indexed with the program counter (PC) of load instructions. Each entry in the RPT corresponds to a specific load instruction; it keeps the Last Block referenced by the instruction and the Last Stride observed in the stream (i.e., the distance of two last addresses accessed by the instruction).

Upon each trigger access (i.e., a cache miss or a prefetch hit), the RPT is searched with the PC of the instruction. If the search results in a miss, it means that no history does exist for the instruction, and hence, no prefetch request can be issued. Under two circumstances, a search may result in a miss: (1) whenever a load instruction is a new one in the execution flow of the program, and ergo, no history has been recorded for it so far, and (2) whenever a load instruction is re-executed after a long time, and the corresponding recorded metadata information has been evicted from the RPT due to conflicts. In such cases when no matching entry does exist in the RPT, a new entry is allocated for the instruction, and possibly a victim entry is evicted. The new entry is tagged with the PC, and the Last Block field of the entry is filled with the referenced address. The Last Stride is also set to zero (an invalid value) as no stride has yet been observed for this stream. However, if searching the RPT results in a hit, it means that there is a recorded history for the instruction. In this case, the recorded history information is checked with the current access to find out whether or not the stream is a stride one. To do so, the difference of the current address and the Last Block is calculated to get the current stride. Then, the current stride is checked against the recorded Last Stride. If they do not match, it is implied that the stream does not exhibit a stride access pattern. However, if they match, it is construed that the stream is a stride one as three consecutive accesses have produced two identical strides. In this case, based on the lookahead of the prefetcher (Section 2.2), several prefetch requests can be issued by consecutively adding the observed stride to the requested address.

For example, if the current address and stride are A and k, respectively, and the lookahead of prefetching is three, prefetch candidates will be{A + k, A + k + k, A + k + k + k}. Finally, regardless of the fact that the stream is stride or not, the corresponding RPT entry is updated: the Last Block is updated with the current address, and the Last Stride takes the value of the current stride.

4. Nonhardware data prefetching

Progress in technology fabrication accompanied by circuit-level and microarchitectural advancements has brought about significant enhancements in the processors' performance over the past decades. Meanwhile, the performance of memory systems has not improved in paste with that of the processors, forming a large gap between the performance of processors and memory systems [13–17]. As a consequence, numerous approaches have been proposed to enhance the execution performance of applications by bridging the processor-memory performance gap. Hardware data prefetching is just one of these approaches. Hardware data prefetching bridges the gap by proactively fetching the data ahead of the cores' requests to eliminate the idle cycles in which the processor is waiting for the response of the memory system. In this section, we briefly review the other approaches that target the same goal (i.e., bridging the processor-memory performance gap) but in other ways.

Multithreading [18] enables the processor to better utilize its computational resources, as stalls in one thread can be overlapped with the execution of other threads [4,19,20]. Multithreading, however, only improves throughput and does nothing for (or even worsens) the response time [15,21–23], which is crucial for satisfying the strict latency requirements of server applications.

Thread-Based Prefetching techniques [24–28] exploit idle thread contexts or distinct pre-execution hardware to drive helper threads that try to overlap the cache misses with speculative execution. Such helper threads, formed either by the hardware or by the compiler, execute a piece of code that prefetches for the main thread. Nonetheless, the additional threads and fetch/execution bandwidth may not be available when the processor is fully utilized. The abundant request-level parallelism of server applications [22–24] makes such schemes ineffective in that the helper threads need to compete with the main threads for the hardware context.

Runahead Execution [29–31] makes the execution resources of a core that would otherwise be stalled on an off-chip cache miss to go ahead of the stalled execution in an attempt to discover additional load misses. Similarly, Branch Prediction Directed Prefetching [5] utilizes the branch predictor to run in advance of the executing program, thereby prefetching load instructions along the expected future path. Such approaches, nevertheless, are constrained by the accuracy of the branch predictor and can cover simply a portion of the miss latency, since the runahead thread/branch predictor may not be capable of executing far ahead in advance to completely hide a cache miss. Moreover, these approaches can only prefetch independent cache misses [32] and may not be effective for many of the server workloads, e.g., OLTP and Web applications, that are characterized by long chains of dependent memory accesses [2,33].

On the software side, there are efforts to restructure programs to boost chip-level Data Sharing and Data Reuse [34,35] in order to decrease off-chip accesses. While these techniques are useful for workloads with modest datasets, they fall short of efficiency for big-data server workloads, where the multigigabyte working sets of workloads dwarf the few megabytes of on-chip cache capacity. The ever growing datasets of server workloads make such approaches unscalable. Software Prefetching techniques [36–41] profile the program code and insert prefetch instructions to eliminate cache misses. While these techniques are shown to be beneficial for small benchmarks, they usually require significant programmer effort to produce optimized code to generate timely prefetch requests.

Memory-Side Prefetching techniques [42–44] place the hardware for data prefetching near DRAM, for the sake of saving precious SRAM budget. In such approaches (e.g., [43]), prefetching is performed by a user thread running near the DRAM, and prefetched pieces of data are sent to the on-chip caches. Unfortunately, such techniques lose the predictability of core requests [45] and are incapable of performing cache-level optimizations (e.g., avoiding cache pollution [7]).

5. Conclusion

Data cache misses are a major source of performance degradation in applications. Data prefetching is a widely-used technique for reducing the number of data cache misses or their negative effects. Data prefetchers usually benefit from correlations and localities among data accesses to predict future memory references. As there exist several types of correlations among data accesses, there are several types of data prefetchers. In this book, we introduce several important classes of data prefetchers and highlight their strengths and weaknesses.

References

[1] K.R. Lee, Impacts of Information Technology on Society in the New Century, 2001. https://www.zurich.ibm.com/pdf/news/Konsbruck.pdf.

[2] G.E. Moore, Cramming more components onto integrated circuits, Electronics 38 (8) (1965) 114–117.

[3] D. Geer, Chip makers turn to multicore processors, Computertomographie 38 (5) (2005) 11–13.

[4] J.-L. Baer, T.-F. Chen, An effective on-chip preloading scheme to reduce data access penalty, in: *Proceedings of the ACM/IEEEConference on Supercomputing*, 1991.

[5] F. Dahlgren, P. Stenstrom, Hardware-based stride and sequential prefetching in shared-memory multiprocessors, in: *Proceedings of the International Symposium on High-Performance Computer Architecture (HPCA)*, 1995.

[6] P. Lotfi-Kamran, H. Sarbazi-Azad, M. Bakhshalipour, Domino temporal data prefetcher, in: *Proceedings of the International Symposium on High-Performance Computer Architecture (HPCA)*, 2018.

[7] M. Shakerinava, P. Lotfi-Kamran, H. Sarbazi-Azad, M. Bakhshalipour, Bingo spatial data prefetcher, in: *Proceedings of the International Symposium on High-Performance Computer Architecture (HPCA)*, 2019.

[8] J. Kim, P. Sharma, R. Panda, P. Gratz, D. Jimenez, D. Kadjo, B-fetch: branch prediction directed prefetching for Chip-multiprocessors, in: *Proceedings of the International Symposium on Microarchitecture (MICRO)*, 2014.

[9] Z. Fang, A. Zhai, P. Yew, S. Mehta, Multi-stage coordinated prefetching for present-day processors, in: *Proceedings of the International Conference on Supercomputing (ICS)*, 2014.

[10] O. Mutlu, H. Kim, Y.N. Patt, S. Srinath, Feedback directed prefetching: improving the performance and bandwidth-efficiency of hardware prefetchers, in: *Proceedings of the International Symposium on High PerformanceComputer Architecture (HPCA)*, 2007.

[11] S.H. Pugsley, P.V. Gratz, A.L. Narasimha Reddy, C. Wilkerson, Z. Chishti, J. Kim, Path confidence based lookahead prefetching, in: *Proceedings of the International Symposium on Microarchitecture (MICRO)*, 2016.

[12] P. Lotfi-Kamran, A. Mazloumi, F. Samandi, M. Naderan-Tahan, M. Modarressi, S.-A. Hamid, M. Bakhshalipour, Fast data delivery for many-Core processors, IEEE Trans. Comput 67 (10) (2018) 1416–1429.

[13] O. Mutlu, C.J. Lee, Y.N. Patt, E. Ebrahimi, Coordinated control of multiple prefetchers in multi-Core systems, in: *Proceedings of the International Symposium on Microarchitecture (MICRO)*, 2009.

[14] C.J. Lee, O. Mutlu, Y.N. Patt, E. Ebrahimi, Fairness via source throttling: A configurable and high-performance fairness substrate for multi-Core memory systems, in: *Proceedings of the International Conference on Architectural Support for Programming Languages and Operating Systems (ASPLOS)*, 2010.

[15] O. Dongsu Han, M.H.-B. Mutlu, Y. Kim, ATLAS: A scalable and high-performance scheduling algorithm for multiple memory controllers, in: *Proceedings of the International Symposium on High-Performance Computer Architecture (HPCA)*, 2010.

[16] J. Steve Dodson, J. Fields, H.Q. Le, B. Sinharoy, J.M. Tendler, POWER4 system microarchitecture, IBM J. Res. Develop. 46 (1) (2002) 5–25.

[17] J. Doweck, Inside Intel®Core Microarchitecture, in: *IEEE Hot Chips Symposium (HCS)*, 2006.

[18] P. Conway, B. Hughes, The AMD Opteron Northbridge architecture, IEEE Micro 27 (2) (2007) 10–21.

[19] T. Horel, G. Lauterbach, UltraSPARC-III: designing third-generation 64-bit performance, IEEE Micro 19 (3) (1999) 73–85.

[20] S. Sair, B. Calder, T. Sherwood, Predictor-directed stream buffers, in: *Proceedings of the International Symposium on Microarchitecture (MICRO)*, 2000.

[21] M. Inaba, K. Hiraki, Y. Ishii, Access map pattern matching for data cache prefetch, in: *Proceedings of the International Conferenceon Supercomputing (ICS)*, 2009.

[22] T. Sherwood, B. Calder, S. Sair, A decoupled predictor-directed stream prefetching architecture, IEEE Trans. Comput 53 (2003) 260–276.

[23] N.P. Jouppi, Improving direct-mapped cache performance by the addition of a small fully-associative cache and prefetch buffers, in: *Proceedings of the International Symposium on Computer Architecture (ISCA)*, 1990.

[24] S. Palacharla, R.E. Kessler, Evaluating stream buffers as a secondary cache replacement, in: *Proceedings of the International Symposium on Computer Architecture (ISCA)*, 1994.

[25] C. Zhang, S.A. McKee, Hardware-only stream prefetching and dynamic access ordering, in: *Proceedings of the International Conference on Supercomputing (ICS)*, 2000.

[26] L. Spracklen, S. Kadambi, Y. Chou, S.G. Abraham, S. Iacobovici, Effective stream-based and execution-based data prefetching, in: *Proceedings of the International Conference on Supercomputing (ICS)*, 2004.

[27] D.J. DeWitt, M.D. Hill, D.A. Wood, A. Ailamaki, DBMSs on a modern processor: where does time go? in: *Proceedings of the International Conference on Very Large Data Bases (VLDB)*, 1999.

[28] T. Diep, M. Annavaram, B. Hirano, H. Eri, H. Nueckel, J.P. Shen, R.A. Hankins, Scaling and characterizing database workloads: bridging the gap between research and practice, in: *Proceedings of the International Symposium on Microarchitecture (MICRO)*, 2003.

[29] J.-L. Larriba-Pey, Z. Zhang, J. Torrellas, P. Trancoso, The memory performance of DSS commercial workloads in shared-memory multiprocessors, in: *Proceedings of the International Symposium on High-Performance Computer Architecture (HPCA)*, 1997.

[30] W.A. Wulf, S. A., McKee, "hitting the memory wall: implications of the obvious," ACM SIGARCH Comput. Architect. News 23 (1) (1995) 20–24.

[31] M. Nemirovsky, D.M. Tullsen, Multithreading Architecture, first edition, Morgan & Claypool Publishers, 2013.

[32] H. Akkary, M.A. Driscoll, A dynamic multithreading processor, in: *Proceedings of the International Symposium on Microarchitecture (MICRO)*, 1998.

[33] J. Wu, C.-C. Tsai, J. Yang, H. Cui, Stable deterministic multithreading through schedule Memoization, in: *Proceedings of the USENIX Conference on Operating Systems Design and Implementation (OSDI)*, 2010.

[34] C.I. Rodrigues, S.S. Baghsorkhi, S.S. Stone, D.B. Kirk, W.-M.W. Hwu, S. Ryoo, Optimization principles and application performance evaluation of a multithreaded GPU using CUDA, in: *Proceedings of the Symposium on Principles and Practice of Parallel Programming (PPoPP)*, 2008.

[35] L.A. Barroso, S.J. Eggers, K. Gharachorloo, H.M. Levy, S.S. Parekh, J.L. Lo, An analysis of database workload performance on simultaneous multithreaded processors, in: *Proceedings of the International Symposium on Computer Architecture (ISCA)*, 1998.

[36] B. Grot, M. Ferdman, S. Volos, O. Kocberber, J. Picorel, A. Adileh, D. Jevdjic, S. Idgunji, E. Ozer, B. Falsafi, P. Lotfi-Kamran, Scale-Out Processors, in: *Proceedings of the International Symposium on Computer Architecture (ISCA)*, 2012.

[37] M. Bakhshalipour, B. Khodabandeloo, P. Lotfi-Kamran, H. Sarbazi-Azad, P. Esmaili-Dokht, Scale-out Processors & Energy Efficiency, arXiv (2018).

[38] D.M. Tullsen, H. Wang, J.P. Shen, J.D. Collins, Dynamic speculative precomputation, in: *Proceedings of the International Symposium on Microarchitecture (MICRO)*, 2001.

[39] H. Wang, D.M. Tullsen, C. Hughes, Y.-F. Lee, D. Lavery, J.P. Shen, J.D. Collins, Speculative precomputation: long-range prefetching of delinquent loads, in: *Proceedings of the International Symposium on Computer Architecture (ISCA)*, 2001.

[40] I. Ganusov, M. Burtscher, Future execution: A prefetching mechanism that uses multiple cores to speed up single threads, ACM Trans. Architec. Code Optimiz. 3 (4) (2006) 424–449.

[41] S. Swanson, D.M. Tullsen, M. Kamruzzaman, Inter-Core prefetching for multicore processors using migrating helper threads, in: *Proceedings of the International Conference on Architectural Support for Programming Languages and Operating Systems (ASPLOS)*, 2011.

[42] C. Jung, D. Lim, Y. Solihin, J. Lee, Prefetching with helper threads for loosely coupled multiprocessor systems, IEEE Trans. Parallel Distrib. Syst. 20 (9) (2009) 1309–1324.

[43] A. Adileh, O. Kocberber, S. Volos, M. Alisafaee, D. Jevdjic, C. Kaynak, A. Daniel Popescu, A. Ailamaki, B. Falsafi, M. Ferdman, Clearing the clouds: A study of emerging scale-out workloads on modern hardware, in: *Proceedings of the International Conference on Architectural Support for Programming Languages and Operating Systems (ASPLOS)*, 2012.

[44] J. Dundas, T. Mudge, Improving data cache performance by pre-executing instructions under a cache miss, in: *Proceedings of the International Conference on Supercomputing (ICS)*, 1997.

[45] H. Kim, Y.N. Patt, O. Mutlu, Techniques for efficient processing in Runahead execution engines, in: *Proceedings of the International Symposium on Computer Architecture (ISCA)*, 2005.

Further reading

[46] I. Pandis, R. Johnson, N.G. Mancheril, A. Ailamaki, B. Falsafi, N. Hardavellas, Database servers on Chip multiprocessors: limitations and opportunities, in: *Proceedings of the Biennial Conference on Innovative Data Systems Research (CIDR)*, 2007.

[47] J. Stark, C. Wilkerson, Y.N. Patt, O. Mutlu, Runahead execution: an alternative to very large instruction windows for out-of-order processors, in: *Proceedings of the International Symposium on High-Performance Computer Architecture (HPCA)*, 2003.

[48] E. Khubaib, O.M. Ebrahimi, Y.N. Patt, M. Hashemi, Accelerating dependent cache misses with an enhanced memory controller, in: *Proceedings of the International Symposium on Computer Architecture (ISCA)*, 2016.

[49] K. Gharachorloo, S.V. Adve, L.A. Barroso, P. Ranganathan, Performance of database workloads on shared-memory systems with out-of-order processors, in: *Proceedings of the International Conference on Architectural Support for Programming Languages and Operating Systems (ASPLOS)*, 1998.

[50] R. Johnson, N. Hardavellas, I. Pandis, N.G. Mancheril, S. Harizopoulos, K. Sabirli, A. Ailamaki, B. Falsafi, To Share or Not to Share? in: *Proceedings of the International Conference on Very Large Data Bases (VLDB)*, 2007.

[51] J.R. Larus, M. Parkes, Using cohort-scheduling to enhance server performance, in: *Proceedings of the General Track of the Annual Conference on USENIX Annual Technical Conference (ATEC)*, 2002.

[52] A. Ailamaki, P.B. Gibbons, T.C. Mowry, S. Chen, Improving hash join performance through prefetching, ACM Trans. Database Syst. 32 (3) (2007). p. Article 17.

[53] T.M. Chilimbi, M. Hirzel, Dynamic hot data stream prefetching for general-purpose programs, in: *Proceedings of the ACM SIGPLAN Conference on Programming Language Design and Implementation (PLDI)*, 2002.

[54] C.-K. Luk, T.C. Mowry, Compiler-based prefetching for recursive data structures, in: *Proceedings of theInternational Conference on Architectural Support for Programming Languages and Operating Systems (ASPLOS)*, 1996.
[55] A. Roth, G.S. Sohi, Effective jump-pointer prefetching for linked data structures, in: *Proceedings of the International Symposium on Computer Architecture (ISCA)*, 1999.
[56] A. Mirhosseini, S.B. Ehsani, H. Sarbazi-Azad, M. Drumond, B. Falsafi, R. Ausavarungnirun, O. Mutlu, M. Sadrosadati, LTRF: enabling high-capacity register files for GPUs via hardware/software cooperative register prefetching, in: *Proceedings of the International Conference on Architectural Support for Programming Languages and Operating Systems (ASPLOS)*, 2018.
[57] B. Calder, D.M. Tullsen, W. Zhang, A self-repairing prefetcher in an event-driven dynamic optimization framework, in: *Proceedings of the International Symposium on Code Generation and Optimization (CGO)*, 2006.
[58] C.J. Hughes, S.V. Adve, Memory-side prefetching for linked data structures for processor-in-memory systems, J. Parallel Distrib. Comput. 65 (4) (2005) 448–463.
[59] J. Lee, J. Torrellas, Y. Solihin, Using a user-level memory thread for correlation prefetching, in: *Proceedings of the International Symposium on Computer Architecture (ISCA)*, 2002.
[60] J. Kotra, E. Kultursay, M. Kandemir, C.R. Das, A. Sivasubramaniam, P. Yedlapalli, Meeting midway: improving CMP performance with memory-side prefetching, in: *Proceedings of the International Conference on Parallel Architectures and Compilation Techniques (PACT)*, 2013.
[61] S. Mittal, A survey of recent prefetching techniques for processor caches, ACM Compu. Surveys 49 (2) (2016). p. 35:1–35:35.

About the authors

Pejman Lotfi-Kamran is an associate professor and the head of the school of computer science and the director of Turin Cloud Services at Institute for Research in Fundamental Sciences (IPM). His research interests include computer architecture, computer systems, approximate computing, and cloud computing. His work on scale-out server processor design lays the foundation for Cavium ThunderX. Lotfi-Kamran has a Ph.D. in computer science from the École Polytechnique Fédérale de Lausanne (EPFL). He received his MS and BS in computer engineering from the University of Tehran. He is a member of the IEEE and the ACM.

Hamid Sarbazi-Azad is currently a professor of computer science and engineering at the Sharif University of Technology, Tehran, Iran. His research interests include high-performance computer/memory architectures, NoCs and SoCs, parallel and distributed systems, social networks, and storage systems, on which he has published over 400 refereed papers. He received Khwarizmi International Award in 2006, TWAS Young Scientist Award in engineering sciences in 2007, Sharif University Distinguished Researcher awards in the years 2004, 2007, 2008, 2010, and 2013, the Iranian Ministry of Communication and Information Technology's award for contribution to IT research and education in 2014, and distinguished book author of the Sharif University of Technology in 2015 and 2020. Dr Sarbazi-Azad is now an associate editor of ACM Computing Surveys, IEEE Computer Architecture Letters, and Elsevier's Computers and Electrical Engineering.

CHAPTER TWO

Spatial prefetching

Pejman Lotfi-Kamran[a] and Hamid Sarbazi-Azad[b]
[a]School of Computer Science, Institute for Research in Fundamental Sciences (IPM), Tehran, Iran
[b]Sharif University of Technology, and Institute for Research in Fundamental Sciences (IPM), Tehran, Iran

Contents

Abstract

Many applications extensively use data objects with a regular and fixed layout, which leads to the recurrence of access patterns over memory regions. Spatial data prefetching techniques exploit this phenomenon to prefetch future memory references and hide their long latency. Spatial prefetchers are particularly of interest because they usually only need a small storage budget. In this chapter, we introduce the concept of spatial prefetching and present two instances of spatial data prefetchers, SMS and VLDP.

1. Introduction

Data prefetchers need to predict future memory accesses of a processor while an application is running. For this purpose, they usually rely on a recurring pattern or correlation among data accesses. The more the pattern or a particular correlation occurs in the data accesses of an application while running on a processor, the better the data prefetcher can predict future memory references.

Spatial data prefetchers predict future memory accesses by relying on spatial address correlation. Spatial address correlation is the similarity of data access patterns among multiple *regions* of memory. Access patterns demonstrate spatial correlation because applications use data objects with a regular and fixed layout, and accesses reoccur when data structures are traversed [1]. For example, applications may use an array of structs while the struct has multiple elements. Likely, the pattern of accesses to different structs is similar (or sometimes identical). Fig. 1 shows the access patterns of several regions of memory. The similarity of access patterns is visible in this figure. Such similarity of access patterns to regions of memory (in this case, the memory regions related to structs) is referred to as spatial correlation.

There are several spatial data prefetchers [1–10]. These prefetchers divide the memory address space into fixed-size sections, named *Spatial Regions,* and learn the memory access patterns over these sections. The learned access patterns are then used for prefetching future memory references when the application touches the *same* or *similar Spatial Regions.* In this chapter, we talk about the machisms by which spatial prefetchers learn access patterns in *Spatial Regions,* and how they apply what they learned to prefetch for similar *Spatial Regions* in the future.

One of the strengths of spatial data prefetchers is that they impose low area overhead because they store *offsets* (i.e., the distance of a block address from the beginning of a *Spatial Region)* or *deltas* (i.e., the distance of two consecutive accesses that fall into a *Spatial Region)* as their metadata information, and not complete addresses. Another equally remarkable strength of spatial data prefetchers is their ability to eliminate compulsory cache misses [11]. Compulsory misses are those cache misses for which the address is seen by the cache for the first time. These misses are also known as cold start misses or first reference misses. As the cache sees these addresses for the first time, such data accesses usually miss in the cache. Compulsory cache misses, hence, are a major source of performance degradation in important classes of applications, e.g., scan-dominated workloads, where scanning large volumes of data

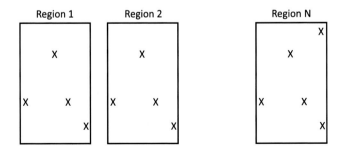

access = X

Fig. 1 The access patterns of several regions of memory.

produces a bulk of unseen memory accesses that cannot be captured by caches [1]. By utilizing the pattern that was observed in a past *Spatial Region* to a new unobserved *Spatial Region*, spatial prefetchers can alleviate the compulsory cache misses, significantly enhancing system performance.

The critical limitation of spatial data prefetching is its ineptitude in predicting pointer-chasing–caused cache misses [12–16]. As an example, with linked lists, every element is dynamically allocated somewhere in memory, and elements are not necessarily adjacent to each other. Instead, they are linked together using pointers. In general, as dynamic objects can potentially be allocated everywhere in the memory, and hence are not necessarily adjacent to each other, pointer-chasing accesses do not usually exhibit spatial correlation. Instead, they produce bulks of dependent cache misses for which spatial prefetchers can do very little due to a lack of spatial address correlation.

We consider two state-of-the-art spatial prefetching techniques: (1) SPATIAL MEMORY STREAMING [2], and (2) VARIABLE LENGTH DELTA PREFETCHER [4].

2. Spatial memory streaming (SMS)

SMS is a state-of-the-art spatial prefetcher that was proposed and evaluated in the context of server and scientific applications. Whenever a *Spatial Region* is requested for the first time, SMS starts to observe and record accesses to that *Spatial Region* as long as the *Spatial Region* is actively used by the application. Whenever the *Spatial Region* is no longer utilized (i.e., the corresponding blocks of the *Spatial Region* start to be evicted from the cache), SMS stores the information of the observed accesses in its metadata table, named *Pattern History Table (PHT)*.

The information in *PHT* is stored in the form of $<event, pattern>$. The *event* is a piece of information to which the observed access pattern is correlated. That is, it is expected for the stored access pattern to be used whenever *event* reoccurs in the future. SMS empirically chooses $PC + Offset$ of the trigger access (i.e., the PC of the instruction that first accesses the *Spatial Region* combined with the distance of the first requested cache block from the beginning of the *Spatial Region*) as the *event* to which the access patterns are correlated. Doing so, whenever a $PC + Offset$ is reoccurred, the correlated access pattern history is used for issuing prefetch requests. The *pattern* is the history of accesses that happen in every *Spatial Region*. SMS encodes the patterns of the accesses as a *bit vector*. In this manner, for every cache block in a *Spatial Region*, a bit is stored, indicating whether the block has

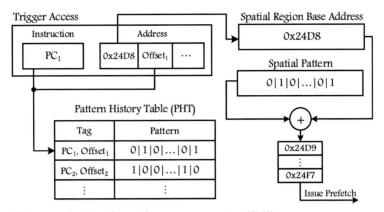

Fig. 2 The organization of spatial memory streaming (SMS).

been used during the latest usage of the *Spatial Region* ('1') or not ('0'). Therefore, whenever a *pattern* is going to be used, prefetch requests are issued only for blocks whose corresponding bit in the stored *pattern* is '1.' Fig. 2 shows the hardware realization of SMS.

2.1 Example

Let's assume that SMS tracks *Spatial Regions* of size 256 bytes and the cache is direct-mapped with the size of 512 bytes while block size is 64 bytes. In this example, the cache contains eight blocks, and each region is 4-block long.

We consider the following sequence of memory accesses in this example:

67, 200, 580, 1346, 1490, 1310

In the beginning, both the cache and the prefetcher states are empty. Address 67 is not in the cache and maps to its second block. Moreover, it is the first access to a new *Spatial Region*. Hence, it triggers SMS. SMS records the PC of the access (let say PC = 0x001F54) and the offset 64, as this address is the second block of the *Spatial Region*.

The second memory access is Address 200. This address is also not in the cache and maps to its fourth block. Moreover, Address 200 is from the same *Spatial Region* as Address 67, as such, the offset 192 will be recorded.

The next memory access is Address 580. This address is also not in the cache and maps to its second block. Consequently, Address 67 will be replaced with this address in the cache. As one of the blocks of an active *Spatial Region* is replaced in the cache, it indicates that the *Spatial Region* is no longer active.

Consequently, {PC + Offset, Bit Pattern} (in this case, {0x001F54+64, {0,1,0,1}} will be recorded. Address 580 also starts a new *Spatial Region*. For brevity, we do not mention what SMS records for this *Spatial Region*. The next memory access is Address 1346. This address is not in the cache and maps to the fifth block of the cache. This access is also the beginning of a new *Spatial Region*. For brevity, we do not mention what SMS does to track accesses in this *Spatial Region*. SMS also checks if this access matches any of the recorded PC + Offsets. Assume this memory access is from the same load instruction as the prior memory accesses. Consequently, the PCs are identical. As the offset in the *Spatial Region* is 64, the offset also matches. Therefore, SMS prefetches Pattern {0,1,0,1} in the new *Spatial Region* (i.e., only Block Address 1472 as Block Address 1344 was the trigger).

The next address is 1490. As previously, Block Address 1472 is prefetched, Address 1490 is already in the cache, and hence, a hit. The last address is 1310 which is from the same *Spatial Region* as the previous memory access. However, this address is not in the cache because the previous pattern that SMS used for prefetching did not have this block, and consequently, SMS did not prefetch this cache block.

3. Variable length delta prefetcher (VLDP)

VLDP is a recent state-of-the-art spatial data prefetcher that relies on the similarity of *delta* patterns among *Spatial Regions* of memory. VLDP records the distance between consecutive accesses that fall into *Spatial Regions* and uses them to predict future misses. The key innovation of VLDP is the deployment of *multiple* prediction tables for predicting delta patterns. VLDP employs several history tables where each table keeps the metadata based on a specific length of the input history.

Fig. 3 shows the metadata organization of VLDP. The three major components are *Delta History Buffer (DHB), Delta Prediction Table (DPT),* and *Offset Prediction Table (OPT).* DHB is a small table that records the delta history of *currently-active Spatial Regions.* Each entry in DHB is associated with an active *Spatial Region* and contains details like the *Last Referenced Block.* These details are used to index OPT and DPTs for issuing prefetch requests.

DPT is a set of key-value pairs that correlates a delta sequence to the next expected delta. VLDP benefits from multiple DPTs where each DPT records the history with a different length of the input. DPT-*i* associates

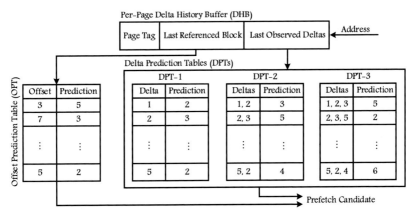

Fig. 3 The organization of variable length delta prefetcher (VLDP).

a sequence of i deltas to the next expected delta. For example, if the last three deltas in a *Spatial Region* are d_3, d_2, and d_1 (d_1 is the most recent delta), *DPT-2* stores $[<d_3, d_2> \rightarrow d_1]$, while *DPT-1* records $[<d_2> \rightarrow d_1]$. While looking up the *DPTs,* if several of them offer a prediction, the prediction of the table with the *longest* sequence of deltas is used, because predictions that are made based on longer inputs are expected to be more accurate [1]. This way, VLDP eliminates wrong predictions that are made by short inputs, enhancing both accuracy and miss coverage of the prefetcher.

OPT is another metadata table of VLDP, that is indexed using the *offset* (and not delta) of the first access to a *Spatial Region.* Merely relying on deltas for prefetching requires the prefetcher to observe at least the first two accesses to a *Spatial Region* before issuing prefetch requests; however, there are many *sparse Spatial Regions* in which a few, say, two, of the blocks are used by the application. Therefore, waiting for two accesses before starting the prefetching may divest the prefetcher of issuing enough prefetch requests when the application operates on a significant number of sparse *Spatial Regions.* Employing *OPT* enables VLDP to start prefetching immediately after the first access to *Spatial Regions. OPT* associates the offset of the first access of a *Spatial Region* to the next expected delta. After the first access to a *Spatial Region, OPT* is looked up using the offset of the access, and the output of the table is used for issuing a prefetch request. For the rest of the accesses to the *Spatial Region* (i.e., second access onward), VLDP uses only *DPTs.*

Even though VLDP relies on prediction tables with a single next expected delta, it is still able to offer a prefetching lookahead larger than one, using the proposed *multi-degree* prefetching mechanism. In the *multi-degree* mode, upon predicting the next delta in a *Spatial Region*, VLDP *uses the prediction as an input for DPTs* to make more predictions.

3.1 Example

Similar to SMS' example, let's assume that VLDP tracks spatial regions of size 256 bytes, and the cache is direct-mapped with 512-byte size and block size is 64 bytes. In this example, the cache contains eight blocks, and each region is 4-block long.

We consider the same sequence of addresses that we used for SMS:
67, 200, 580, 1346, 1490, 1310

In the beginning, both the cache and the prefetcher states are empty. Address 67 is not in the cache and maps to its second block. Moreover, as this is the first address of the *Spatial Region*, there is no entry associated with it in the Delta History Buffer (DHB). So the following entry will be added to DHB (Note that offsets and deltas are in terms of cache blocks and not bytes) (Table 1).

The second memory access is Address 200. This address is also not in the cache and maps to its fourth block. Moreover, Address 200 is from the same *Spatial Region* as Address 67, as such, the DHB entry will be updated (Table 2).

Moreover, as this is the second access to this *Spatial Region*, an entry will be added to the offset prediction table (OPT) (Table 3).

Table 1 Delta history buffer (DHB)—67.

Spatial region no.	Last address	Last 4 deltas	...
0	1		...

Table 2 Delta history buffer (DHB)—67, 200.

Spatial region no.	Last address	Last 4 deltas	...
0	3	2	...

Table 3 Offset prediction table (OPT)—67, 200.

Offset	Delta
1	2

The next memory access is Address 580. This address is also not in the cache and maps to its second block. Consequently, Address 67 will be replaced with this address in the cache. As this address is from a different *Spatial Region*, an entry will be created for it in DHB (Table 4).

Moreover, as this is the first access to a *Spatial Region*, it qualifies for a prefetching using OPT. The offset of this access from the beginning of the *Spatial Region* is 1 and because an entry with this offset exists in OPT, VLDP prefetches $BlockAddress(580) + 2 \times BlockSize = 576 + 128 = 704$.

The next memory access is Address 1347. This address is not in the cache and maps to the fifth block of the cache. This access is also the beginning of a new spatial region. The content of DHB is shown in Table 5.

Moreover, as this is the first access to the *Spatial Region*, it qualifies for a prefetching using OPT. The offset of this access from the beginning of the *Spatial Region* is 1. Using OPT, VLDP prefetches $BlockAddress(1347) + 2 \times BlockSize = 1344 + 128 = 1472$.

The next address is 1490. As previously, Block Address 1472 is prefetched, Address 1490 is already in the cache, and hence, a hit. After 1472, the content of DHB is shown in Table 6.

Table 4 Delta history buffer (DHB)—67, 200, 580.

Spatial region no.	Last address	Last 4 deltas	...
0	3	2	...
2	1		...

Table 5 Delta history buffer (DHB)—67, 200, 580, 1347.

Spatial region no.	Last address	Last 4 deltas	...
0	3	2	...
2	1		...
5	1		...

Table 6 Delta history buffer (DHB)—67, 200, 580, 1347, 1490.

Spatial region no.	Last address	Last 4 deltas	...
0	3	2	...
2	1		...
5	3	2	...

Table 7 Delta history buffer (DHB)—67, 200, 580, 1347, 1490, 1310.

Spatial region no.	Last address	Last 4 deltas	...
0	3	2	...
2	1		...
5	0	2, −3	...

The last address is 1310, which is from the same spatial region. However, this address is not in the cache because VLDP did not prefetch it. The content of DHB after this memory access is shown in Table 7.

Moreover, as we have more than one delta in a row of the DHB, we may add entries to the Delta Prediction Table (DPT). In this case as there are just two deltas, we may add "2 ==> −3" to DPT-1. For future memory accesses, this entry can be used for prefetching.

4. Summary

Spatial prefetching has been proposed and developed to capture the similarity of access patterns among memory pages (e.g., if a program visits locations $\{A,B,C,D\}$ of Page X, it is probable that it visits locations $\{A, B,C, D\}$ of other pages as well). Spatial prefetching works because applications use data objects with a regular and fixed layout, and accesses reoccur when data structures are traversed. Spatial prefetching is appealing since it imposes low storage overhead to the system, paving the way for its adoption in future systems.

References

[1] M. Bakhshalipour, M. Shakerinava, P. Lotfi-Kamran, H. Sarbazi-Azad, Bingo spatial data prefetcher, in: Proceedings of the International Symposium on High-Performance Computer Architecture (HPCA), 2019, pp. 399–411.
[2] S. Somogyi, T.F. Wenisch, A. Ailamaki, B. Falsafi, A. Moshovos, Spatial memory streaming, in: Proceedings of the International Symposium on Computer Architecture (ISCA), 2006, pp. 252–263.
[3] K.J. Nesbit, J.E. Smith, Data cache prefetching using a global history buffer, in: Proceedings of the International Symposium on High Performance Computer Architecture (HPCA), 2004, p. 96.
[4] M. Shevgoor, S. Koladiya, R. Balasubramonian, C. Wilkerson, S.H. Pugsley, Z. Chishti, Efficiently prefetching complex address patterns, in: Proceedings of the International Symposium on Microarchitecture (MICRO), 2015, pp. 141–152.
[5] K.J. Nesbit, A.S. Dhodapkar, J.E. Smith, AC/DC: an adaptive data cache Prefetcher, in: Proceedings of the International Conference on Parallel Architectures and Compilation Techniques (PACT), 2004, pp. 135–145.

[6] S. Kumar, C. Wilkerson, Exploiting spatial locality in data caches using spatial footprints, in: Proceedings of the International Symposium on Computer Architecture (ISCA), 1998, pp. 357–368.

[7] C.F. Chen, S.-H. Yang, B. Falsafi, A. Moshovos, Accurate and complexity-effective spatial pattern prediction, in: Proceedings of the International Symposium on High Performance Computer Architecture (HPCA), 2004, pp. 276–287.

[8] J.F. Cantin, M.H. Lipasti, J.E. Smith, Stealth prefetching, in: Proceedings of the International Conference on Architectural Support for Programming Languages and Operating Systems (ASPLOS), 2006, pp. 274–282.

[9] J. Kim, S.H. Pugsley, P.V. Gratz, A.L.N. Reddy, C. Wilkerson, Z. Chishti, Path confidence based Lookahead prefetching, in: Proceedings of the International Symposium on Microarchitecture (MICRO), 2016. pp. 60:1–60:12.

[10] M. Bakhshalipour, M. Shakerinava, P. Lotfi-Kamran, H. Sarbazi-Azad, Accurately and maximally prefetching spatial data access patterns with bingo, The Third Data Prefetching Championship (2019).

[11] D.A. Patterson, J.L. Hennessy, Computer Organization and Design, Fifth Edition: The Hardware/Software Interface, fifth ed., Morgan Kaufmann Publishers Inc., San Francisco, CA, USA, 2013.

[12] M. Bakhshalipour, P. Lotfi-Kamran, H. Sarbazi-Azad, An efficient temporal data Prefetcher for L1 caches, IEEE Computer Architecture Letters (CAL) vol. 16 (2) (2017) 99–102.

[13] M. Bakhshalipour, P. Lotfi-Kamran, H. Sarbazi-Azad, in: Domino Temporal Data Prefetcher, Proceedings of the International Symposium on High-Performance Computer Architecture (HPCA), IEEE, 2018, pp. 131–142.

[14] S. Somogyi, T.F. Wenisch, A. Ailamaki, B. Falsafi, Spatio-temporal memory streaming, in: Proceedings of the International Symposium on Computer Architecture (ISCA), 2009, pp. 69–80.

[15] T.F. Wenisch, M. Ferdman, A. Ailamaki, B. Falsafi, A. Moshovos, Practical off-Chip Meta-data for temporal memory streaming, in: Proceedings of the International Symposium on High Performance Computer Architecture (HPCA), 2009, pp. 79–90.

[16] T.F. Wenisch, S. Somogyi, N. Hardavellas, J. Kim, A. Ailamaki, B. Falsafi, Temporal streaming of shared memory, in: Proceedings of the International Symposium on Computer Architecture (ISCA), 2005, pp. 222–233.

About the authors

Pejman Lotfi-Kamran is an associate professor and the head of the school of computer science and the director of Turin Cloud Services at Institute for Research in Fundamental Sciences (IPM). His research interests include computer architecture, computer systems, approximate computing, and cloud computing. His work on scale-out server processor design lays the foundation for Cavium ThunderX. Lotfi-Kamran has a Ph.D. in computer science from the École Polytechnique Fédérale de Lausanne (EPFL). He received his MS and BS in computer engineering from the University of Tehran. He is a member of the IEEE and the ACM.

Hamid Sarbazi-Azad is currently a professor of computer science and engineering at the Sharif University of Technology, Tehran, Iran. His research interests include high-performance computer/memory architectures, NoCs and SoCs, parallel and distributed systems, social networks, and storage systems, on which he has published over 400 refereed papers. He received Khwarizmi International Award in 2006, TWAS Young Scientist Award in engineering sciences in 2007, Sharif University Distinguished Researcher awards in the years 2004, 2007, 2008, 2010, and 2013, the Iranian Ministry of Communication and Information Technology's award for contribution to IT research and education in 2014, and distinguished book author of the Sharif University of Technology in 2015 and 2020. Dr Sarbazi-Azad is now an associate editor of ACM Computing Surveys, IEEE Computer Architecture Letters, and Elsevier's Computers and Electrical Engineering.

CHAPTER THREE

Temporal prefetching

Pejman Lotfi-Kamran[a] and Hamid Sarbazi-Azad[b]
[a]School of Computer Science, Institute for Research in Fundamental Sciences (IPM), Tehran, Iran
[b]Sharif University of Technology, and Institute for Research in Fundamental Sciences (IPM), Tehran, Iran

Contents

Abstract

Many applications, including big-data server applications, frequently encounter data misses. Consequently, they lose significant performance potential. Fortunately, data accesses of many of these applications follow temporal correlations, which means data accesses repeat over time. Temporal correlations occur because applications usually consist of loops, and hence, the sequence of instructions that constitute the body of a loop repeats many times, leading to data access repetition. Temporal data prefetchers take advantage of temporal correlation to predict and prefetch future memory accesses. In this chapter, we introduce the concept of temporal prefetching and present two instances of temporal data prefetchers, STMS and ISB.

1. Introduction

Temporal prefetching refers to replaying the sequence of past cache misses in order to avert future misses. Temporal prefetching works because in many applications, the sequence of cache misses, which we refer to as a stream, repeats. Temporal data prefetchers [1–16] record the sequence of data misses in the order they appear and use the recorded history for predicting future data misses. Upon a new data miss, they search the history and find a matching entry and replay the sequence of data misses after the match in an attempt to eliminate potential future data misses. A tuned version of temporal prefetching has been implemented in *IBM Blue Gene/Q*, where it is called LIST PREFETCHING [17].

Temporal prefetching is an ideal choice to eliminate long chains of *dependent* cache misses, that are common in pointer-chasing applications (e.g., *OLTP* and *Web*) [5,8,18,19]. A dependent cache miss refers to a memory operation that results in a cache miss and is dependent on data from a prior cache miss. Such misses have a marked effect on the execution performance of applications and impede the processor from making forward progress since both misses are fetched *serially* [8,20]. Because of the lack of strided/spatial correlation among dependent misses, stride and spatial prefetchers are usually unable to prefetch such misses [21]; however, temporal prefetchers, by recording and replaying the sequences of data misses, can prefetch dependent cache misses and result in a significant performance improvement [22,23].

Temporal prefetchers, on the other face of the coin, also have shortcomings. Temporal prefetching techniques exhibit low accuracy as they do not know where streams end. That is, in the foundation of temporal prefetching, there is no wealth of information about *when prefetching should be stopped*; hence, temporal prefetchers continue issuing many prefetch requests until another triggering event occurs, resulting in overprediction.

Moreover, as temporal prefetchers rely on address repetition, they are unable to prevent compulsory misses (unobserved misses) from happening. In other words, they can only prefetch cache misses that at least once have been observed in the past; however, there are many important applications (e.g., *DSS*) in which the majority of cache misses occurs only once during the execution of the application [24,25], for which temporal prefetching can do nothing.

Furthermore, as temporal prefetchers require to store the correlation between addresses, they usually impose large storage overhead (tens of megabytes) that cannot be accommodated on-the-chip next to the cores. Consequently, temporal prefetchers usually place their metadata tables off-the-chip in the main memory. Unfortunately, placing the history information off-the-chip increases the latency of accessing metadata, and more importantly, results in a drastic increase in the off-chip bandwidth consumption for fetching and updating the metadata.

In this chapter, we discuss two state-of-the-art temporal prefetching techniques: (1) SAMPLED TEMPORAL MEMORY STREAMING [4,26], and (2) IRREGULAR STREAM BUFFER [7].

2. Sampled temporal memory streaming (STMS)

STMS is a state-of-the-art temporal data prefetcher that was proposed and evaluated in the context of server and scientific applications. The main observation behind STMS is that *the length of temporal streams widely differs* across programs and across different streams in a particular program, as well, ranging from a couple to hundreds of thousands of cache misses. In order to efficiently store the information of various streams, STMS uses a circular FIFO buffer, named *History Table,* and appends every observed cache miss to its end. This way, the prefetcher is not required to fix a specific predefined length for temporal streams in the metadata organization, that would result in wasting storage for streams shorten than the predefined length or discarding streams longer than it; instead, all streams are stored next to each other in a storage-efficient manner. For locating every address in the *History Table,* STMS uses an auxiliary set-associative structure, named *Index Table.* The *Index Table* stores a *pointer* for every observed miss address to its last occurrence in the *History Table.* Therefore, whenever a cache miss occurs, the prefetcher first looks up the *Index Table* with the missed address and gets the corresponding pointer. Using the pointer, the prefetcher proceeds to the *History Table* and issues prefetch requests for addresses that have followed the missed address in the history.

Fig. 1 shows the metadata organization of STMS, which mainly consists of a *History Table* and an *Index Table.* As both tables require multi-megabyte storage for STMS to have reasonable miss coverage, both tables are placed off-the-chip in the main memory. Consequently, every access to these tables (read or update) should be sent to the main memory and brings/ updates a cache block worth of data. That is, for every stream, STMS needs to wait for two long (serial) memory requests to be sent (one to read the

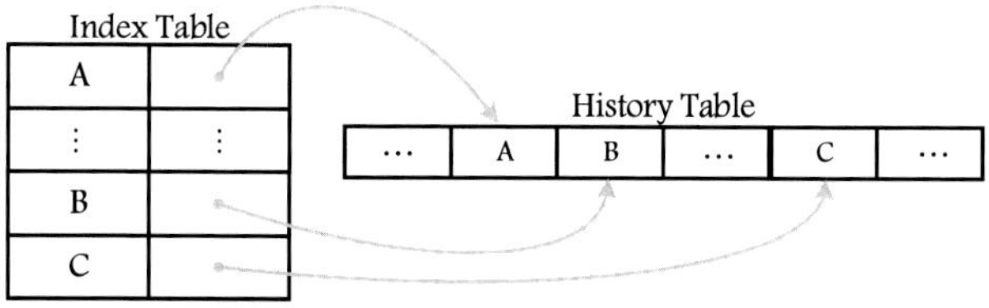

Fig. 1 The organization of sampled temporal memory streaming (STMS).

Index Table and one to read the correct location of the *History Table)* and their responses to come back to the prefetcher before issuing prefetch requests for the stream. The delay of the two off-chip memory accesses, however, is compensated over several prefetch requests of a stream if the stream is long enough.

2.1 Example

Let's assume that the cache is direct-mapped with the size of 512 bytes and the block size is 64 bytes. In this example, the cache contains eight blocks.

We consider the following sequence of addresses to study how STMS works:

    300, 171, 225, 180, 200, 300, 171, 24, 180

In the beginning, both the cache and the prefetcher states are empty. Address 300 is not in the cache and maps to its fifth block. STMS records this address in its circular buffer.

The second memory access is Address 171. This address is also not in the cache and maps to its third block. Moreover, STMS records this address in its circular buffer right after Address 300.

The next memory access is Address 225. This address is also not in the cache and maps to its fourth block. Moreover, STMS records this address in its circular buffer right after Address 171.

The next memory access is Address 180. This address is also not in the cache and maps to its third block. Consequently, Address 171 will be replaced with this address in the cache. Moreover, STMS records this address in its circular buffer right after Address 225.

The next memory access is Address 200. This address is also not in the cache and maps to its fourth block. Consequently, Address 225 will be replaced with this address in the cache. Moreover, STMS records this address in its circular buffer right after Address 180. The state of the circular buffer of STMS after receiving Address 200 is shown in Table 1.

The next memory access is Address 300. This address is in the cache (is a hit). This is the first address that is already seen, and hence, can be found in the STMS circular buffer. Consequently, STMS locates this address in the circular buffer and considers the following addresses as a new stream.

Table 1 Circular buffer.

300	171	225	180	200

STMS sends prefetch requests for addresses in the found stream. If we assume that the prefetching degree is two, Addresses 171, 225 will be prefetched. Address 171 will replace Address 180 in the cache. As Address 225 is already in the cache, no prefetch request will be sent for it. Finally, Address 300 will be appended at the end of the circular buffer.

The next address is 171, which is a hit, thanks to STMS prefetcher. This prefetch hit causes the STMS prefetcher to go ahead in the already found stream and sends a prefetch for Address 180, which will arive to the cache and will replace Address 171. Similar to other addresses, Address 171 will be appended at the end of the circular buffer.

The next address is 24, which is not in the cache and maps to its first block. As this address is seen for the first time and hence is not in the circular buffer, it does not lead to a new stream. Similar to other addresses, Address 24 will be appended at the end of the circular buffer.

The last address is 180, which is in the cache, thanks to the STMS prefetcher.

3. Irregular stream buffer (ISB)

ISB is another state-of-the-art proposal for temporal data prefetching that targets irregular streams of temporally-correlated memory accesses. Unlike STMS that operates on the global miss sequences, ISB attempts to extract temporal correlation among memory references on a per load instruction basis. The key innovation of ISB is the introduction of an extra *indirection* level for storing metadata information. ISB defines a new conceptual address space, named *Structural Address Space (SAS)*, and *maps* the temporally-correlated physical address to this address space in a way that they appear *sequentially*. That is, with this indirection mechanism, physical addresses that are temporally-correlated and used one after another, regardless of their distribution in the *Physical Address Space (PAS)* of memory, become spatially-located and appear one after another in *SAS*. Fig. 2 shows a high-level example of this linearization.

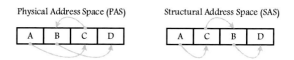

Fig. 2 An example of linearizing scattered temporally-correlated memory references.

ISB utilizes two tables to record a *bidirectional* mapping between address in *PAS* and *SAS*: one table, named *Physical-to-Structural Address Mapping (PSAM)*, records temporally-correlated physical addresses and their mapping information (i.e., to which location in *SAS* they are mapped); the other table, named *Structural-to-Physical Address Mapping (SPAM)*, keeps the *linearized* form of physical addresses in *SAS* and the corresponding mapping information (i.e., which physical addresses are mapped to every structural address). The main purpose of such a linearization is to represent the metadata in a spatially-located manner, paving the way to putting the metadata off-the-chip and *caching* its content in on-chip structures [27]. Like STMS, ISB puts its metadata information off-the-chip to save the precious SRAM storage; however, unlike them, ISB caches the content of its off-chip metadata tables in on-chip structures. Caching the metadata works for ISB as a result of the provided spatial locality. By caching the metadata information, ISB (1) provides faster access to metadata since the caches offer a high hit ratio, and it is not required to proceed to the off-chip memory for every metadata access, and (2) reduces the metadata-induced off-chip bandwidth overhead as many of the metadata manipulations coalesce in the on-chip caches. Fig. 3 shows an overview of the metadata structures of ISB.

Another important contribution of ISB is the synchronization of off-chip metadata manipulations with Translation Lookaside Buffer (TLB) misses. That is, whenever a TLB miss occurs, concurrent with resolving the miss, ISB fetches the corresponding metadata information from the off-chip metadata tables; moreover, whenever a TLB entry is evicted, ISB evicts its corresponding entry from the on-chip metadata structures and updates the off-chip metadata tables. Doing so, ISB ensures that the required metadata is always present in the on-chip structures, significantly hiding the latency of off-chip memory accesses that would otherwise be exposed.

Physical-to-Structural Address Mapping (PSAM)		Structural-to-Physical Address Mapping (SPAM)	
Physical Address	Structural Address	Structural Address	Physical Address
A	m	m, m+1, ...	A, B, ...
⋮	⋮	⋮	⋮
B	m+1	n, n+1, ...	X, Y, ...
X	n	⋮	⋮

Fig. 3 The organization of irregular stream buffer (ISB).

3.1 Example

Let's assume that the cache is direct-mapped with the size of 512 bytes and the block size is 64 bytes. In this example, the cache contains eight blocks.

We consider the same sequence of addresses that we used for STMS prefetcher.

300, 171, 225, 180, 200, 300, 171, 24, 180

Unlike STMS, ISB localizes addresses based on the load instructions that generated them. In this example, we assume that two load instructions generated these addresses, as shown in Table 2. In this example, for brevity, we do not translate addresses and work directly with the original address space.

In the beginning, both the cache and the prefetcher states are empty.

Address 300 is not in the cache and maps to its fifth block. ISB records this address for Load 1.

The second memory access is Address 171. This address is also not in the cache and maps to its third block. Moreover, ISB records this address for Load 1.

The next memory access is Address 225. This address is also not in the cache and maps to its fourth block. Moreover, ISB records this address for Load 2.

The next memory access is Address 180. This address is also not in the cache and maps to its third block. Consequently, Address 171 will be replaced with this address in the cache. Moreover, ISB records this address for Load 1.

The next memory access is Address 200. This address is also not in the cache and maps to its fourth block. Consequently, Address 225 will be replaced with this address in the cache. Moreover, ISB records this address for Load 2. The state recorded for Load 1 and Load 2 after receiving Address 200 is shown in Table 3.

Table 2 The sequence of addresses generated by Load 1 and Load 2.

Load 1:	300	171	180	171	180
Load 2:	225	200	300	24	

Table 3 The state recorded by ISB after receiving 200.

Load 1:	300	171	180
Load 2:	225	200	

The next memory access is Address 300. This address is in the cache (is a hit). This is the first address that is already seen. However, unlike SMS that works on addresses globally, as mentioned, ISB localized addresses based on their load instruction. The observed Address 300 is issued by Load 2 and this is the first time that Load 2 issues this address. Consequently, no prefetch request will be sent. Address 300 will be recorded for Load 2.

The next address is 171, which is a miss. As 171 is already seen by Load 1 instruction, the following addresses will be prefetched (similar to STMS, depending on the prefetch degree, more than one prefetch may be sent). In this case, Address 180 will be prefetched. Address 171 will be recorded as Load 1's state.

The next address is 24, which is not in the cache and maps to its first block. As this address is seen for the first time and hence is not in the ISB states, it does not lead to a prefetch request. Similar to other addresses, Address 24 will be recorded for Load 2.

The last address is 180, which is in the cache, thanks to the ISB prefetcher.

4. Summary

The memory system is a major bottleneck in processors [28–33]. Temporal prefetching has been proposed and developed to capture temporally-correlated access patterns (i.e., the repetition of access patterns in the same order; e.g., if we observe $\{A, B, C, D\}$, then it is likely for $\{B, C, D\}$ to follow $\{A\}$ in the future). Temporal prefetching is well beneficial in the context of pointer-chasing applications, where applications produce bulks of cache misses that exhibit no spatial correlation, but temporal repetition. Temporal prefetchers, however, impose significant overheads to the system, which is still a grave concern in the research literature.

References

[1] D. Joseph, D. Grunwald, Prefetching using Markov predictors, in: *Proceedings of the International Symposium on Computer Architecture (ISCA)*, 1997, pp. 252–263.

[2] Y. Chou, Low-cost epoch-based correlation prefetching for commercial applications, in: *Proceedings of the International Symposium on Microarchitecture (MICRO)*, 2007, pp. 301–313.

[3] K.J. Nesbit, J.E. Smith, Data cache prefetching using a global history buffer, in: *Proceedings of the International Symposium on High Performance Computer Architecture (HPCA)*, 2004, p. 96.

[4] T.F. Wenisch, M. Ferdman, A. Ailamaki, B. Falsafi, A. Moshovos, Practical off-chip Meta-data for temporal memory streaming, in: *Proceedings of the International Symposium on High Performance Computer Architecture (HPCA)*, 2009, pp. 79–90.

[5] T.F. Wenisch, S. Somogyi, N. Hardavellas, J. Kim, A. Ailamaki, B. Falsafi, Temporal streaming of shared memory, in: *Proceedings of the International Symposium on Computer Architecture (ISCA)*, 2005, pp. 222–233.

[6] Y. Solihin, J. Lee, J. Torrellas, Using a user-level memory thread for correlation prefetching, in: *Proceedings of the International Symposium on Computer Architecture (ISCA)*, 2002, pp. 171–182.

[7] A. Jain, C. Lin, Linearizing irregular memory accesses for improved correlated prefetching, in: *Proceedings of the International Symposium on Microarchitecture (MICRO)*, 2013, pp. 247–259.

[8] M. Bakhshalipour, P. Lotfi-Kamran, H. Sarbazi-Azad, in: Domino Temporal Data Prefetcher, *Proceedings of the International Symposium on High-Performance Computer Architecture (HPCA)*, IEEE, 2018, pp. 131–142.

[9] M. Bakhshalipour, P. Lotfi-Kamran, H. Sarbazi-Azad, An efficient temporal data prefetcher for L1 caches, IEEE Comput Archit Lett 16 (2) (2017) 99–102.

[10] M. Bakhshalipour, P. Lotfi-Kamran, A. Mazloumi, F. Samandi, M. Naderan-Tahan, M. Modarressi, H. Sarbazi-Azad, Fast data delivery for many-core processors, IEEE Trans Comput 67 (10) (2018) 1416–1429.

[11] E. Ebrahimi, O. Mutlu, Y.N. Patt, Techniques for bandwidth- efficient prefetching of linked data structures in hybrid prefetching systems, in: *Proceedings of the International Symposium on High- Performance Computer Architecture (HPCA)*, 2009, pp. 7–17.

[12] M. Ferdman, B. Falsafi, Last-touch correlated data streaming, in: *Proceedings of the International Symposium on Performance Analysis of Systems & Software (ISPASS)*, 2007, pp. 105–115.

[13] C.J. Hughes, S.V. Adve, Memory-side prefetching for linked data structures for processor-in-memory systems, J Parallel Distrib Comput 65 (2005) 448–463.

[14] A.-C. Lai, C. Fide, B. Falsafi, Dead-block prediction & dead- block correlating pre-fetchers, in: *Proceedings of the International Symposium on Computer Architecture (ISCA)*, 2001, pp. 144–154.

[15] F. Golshan, M. Bakhshalipour, M. Shakerinava, A. Ansari, P. Lotfi-Kamran, H. Sarbazi-Azad, Harnessing pairwise-correlating data prefetching with runahead meta-data, IEEE Comput Archit Lett 19 (2) (2020) 130–133, https://doi.org/10.1109/LCA.2020.3019343.

[16] A. Roth, G.S. Sohi, Effective jump-pointer prefetching for linked data structures, in: *Proceedings of the International Symposium on Computer Architecture (ISCA)*, 1999, pp. 111–121.

[17] R. Haring, M. Ohmacht, T. Fox, M. Gschwind, D. Satterfield, K. Sugavanam, P. Coteus, P. Heidelberger, M. Blumrich, R. Wisniewski, A. Gara, G. Chiu, P. Boyle, N. Chist, C. Kim, The IBM blue gene/Q compute chip, IEEE Micro 32 (2) (2012) 48–60.

[18] M. Bakhshalipour, S. Tabaeiaghdaei, P. Lotfi-Kamran, H. Sarbazi-Azad, Evaluation of hardware data prefetchers on server processors, ACM Comput Surv 52 (2019). 52:1–52:29.

[19] T.F. Wenisch, M. Ferdman, A. Ailamaki, B. Falsafi, A. Moshovos, Temporal streams in commercial server applications, in: *Proceedings of the IEEE International Symposium on Workload Characterization (IISWC)*, 2008, pp. 99–108.

[20] M. Hashemi, E. Khubaib, O.M. Ebrahimi, Y.N. Patt, Accelerating dependent cache misses with an enhanced memory controller, in: *Proceedings of the International Symposium on Computer Architecture (ISCA)*, 2016, pp. 444–455.

[21] S. Somogyi, T.F. Wenisch, A. Ailamaki, B. Falsafi, Spatio-temporal memory streaming, in: *Proceedings of the International Symposium on Computer Architecture (ISCA)*, 2009, pp. 69–80.

[22] T.M. Chilimbi, On the stability of temporal data reference profiles, in: *Proceedings of the International Conference on Parallel Architectures and Compilation Techniques (PACT)*, 2001, pp. 151–160.

[23] T.M. Chilimbi, M. Hirzel, Dynamic hot data stream prefetching for general-purpose programs, in: *Proceedings of the ACM SIG-PLAN Conference on Programming Language Design and Implementation (PLDI)*, 2002, pp. 199–209.

[24] P. Trancoso, J.-L. Larriba-Pey, Z. Zhang, J. Torrellas, The memory performance of DSS commercial workloads in shared-memory multiprocessors, in: *Proceedings of the International Symposium on High Performance Computer Architecture (HPCA)*, 1997, pp. 250–260.

[25] M. Bakhshalipour, M. Shakerinava, P. Lotfi-Kamran, H. Sarbazi- Azad, Bingo spatial data Prefetcher, in: *Proceedings of the International Symposium on High-Performance Computer Architecture (HPCA)*, 2019, pp. 399–411.

[26] T.F. Wenisch, M. Ferdman, A. Ailamaki, B. Falsafi, A. Moshovos, Making address-correlated prefetching practical, IEEE Micro 30 (2010) 50–59.

[27] I. Burcea, S. Somogyi, A. Moshovos, B. Falsafi, Predictor Virtualization, in: *Proceedings of the International Conference on Architectural Support for Programming Languages and Operating Systems (ASPLOS)*, 2008, pp. 157–167.

[28] M. Ferdman, A. Adileh, O. Kocberber, S. Volos, M. Alisafaee, D. Jevdjic, C. Kaynak, A.D. Popescu, A. Ailamaki, B. Falsafi, Clearing the clouds: a study of emerging scale-out workloads on modern hardware, in: *Proceedings of the International Conference on Architectural Support for Programming Languages and Operating Systems (ASPLOS)*, 2012, pp. 37–48.

[29] M. Ferdman, A. Adileh, O. Kocberber, S. Volos, M. Alisafaee, D. Jevdjic, C. Kaynak, A.D. Popescu, A. Ailamaki, B. Falsafi, Quantifying the mismatch between emerging scale-out applications and modern processors, ACM Trans Comput Syst 30 (2012). 15:1–15:24.

[30] P. Lotfi-Kamran, B. Grot, M. Ferdman, S. Volos, O. Kocberber, J. Picorel, A. Adileh, D. Jevdjic, S. Idgunji, E. Ozer, B. Falsafi, Scale-out processors, in: *Proceedings of the International Symposium on Computer Architecture (ISCA)*, 2012, pp. 500–511.

[31] P. Lotfi-Kamran, B. Grot, B. Falsafi, NOC-out: microarchitecting a scale-out processor, in: *Proceedings of the 45th Annual ACM/IEEE International Symposium on Microarchitecture (MICRO)*, December 2012, pp. 177–187.

[32] B. Grot, D. Hardy, P. Lotfi-Kamran, C. Nicopoulos, Y. Sazeides, B. Falsafi, Optimizing data-center TCO with scale-out processors, IEEE Micro 32 (2012) 1–63.

[33] P. Esmaili-Dokht, M. Bakhshalipour, B. Khodabandeloo, P. Lotfi-Kamran, H. Sarbazi-Azad, Scale-Out Processors & Energy Efficiency, *arXiv preprint arXiv:1808.04864*, 2018.

About the authors

Pejman Lotfi-Kamran is an associate professor and the head of the school of computer science and the director of Turin Cloud Services at Institute for Research in Fundamental Sciences (IPM). His research interests include computer architecture, computer systems, approximate computing, and cloud computing. His work on scale-out server processor design lays the foundation for Cavium ThunderX. Lotfi-Kamran has a Ph.D. in computer science from the École Polytechnique Fédérale de Lausanne (EPFL). He received his MS and BS in computer engineering from the University of Tehran. He is a member of the IEEE and the ACM.

Hamid Sarbazi-Azad is currently a professor of computer science and engineering at the Sharif University of Technology, Tehran, Iran. His research interests include high-performance computer/memory architectures, NoCs and SoCs, parallel and distributed systems, social networks, and storage systems, on which he has published over 400 refereed papers. He received Khwarizmi International Award in 2006, TWAS Young Scientist Award in engineering sciences in 2007, Sharif University Distinguished Researcher awards in the years 2004, 2007, 2008, 2010, and 2013, the Iranian Ministry of Communication and Information Technology's award for contribution to IT research and education in 2014, and distinguished book author of the Sharif University of Technology in 2015 and 2020. Dr Sarbazi-Azad is now an associate editor of ACM Computing Surveys, IEEE Computer Architecture Letters, and Elsevier's Computers and Electrical Engineering.

CHAPTER FOUR

Beyond spatial or temporal prefetching

Pejman Lotfi-Kamran[a] and Hamid Sarbazi-Azad[b]
[a]School of Computer Science, Institute for Research in Fundamental Sciences (IPM), Tehran, Iran
[b]Sharif University of Technology, and Institute for Research in Fundamental Sciences (IPM), Tehran, Iran

Contents

Abstract

Data prefetchers rely on some forms of correlations or repetitive patterns in the data access streams of applications to predict future memory references. However, not all data prefetchers use either spatial correlations or temporal correlations. To introduce the readers to such prefetchers, in this chapter, we present STeMS and BOP. One prefetcher uses both spatial and temporal correlations to benefit from both worlds. The other one uses the concept of offset prefetching to predict future memory accesses.

1. Introduction

There are data prefetching approaches that are not categorized under spatial prefetching or temporal prefetching. Prominently, *spatiotemporal prefetching* and *offset prefetching* are two important class of data prefetchers that do not rely on (solely) one of the spatial access patterns and temporal access patterns.

The motivation of spatiotemporal prefetching is the fact that spatial [1–10] and temporal prefetching techniques [3, 11–25] capture *separate* sub-sets of cache misses, and hence, each omits a considerable portion of cache misses unpredicted. As a considerable fraction of data misses is predictable only by one of the two prefetching techniques, spatiotemporal prefetching tries to combine them in order to reap the benefits of both methods. Another motivation for spatiotemporal prefetching is the fact that the effectiveness of temporal and spatial prefetching techniques varies across applications. As discussed, pointer-chasing applications (e.g., *OLTP*) produce long chains of dependent cache misses that cannot be effectively captured by spatial prefetching but temporal prefetching. On the contrary, scan-dominated applications (e.g., *DSS*) produce a large number of compulsory cache misses that are predictable by spatial prefetchers and not temporal prefetchers.

Offset prefetching [26, 27] is an evolution of stride prefetching, in which, the prefetcher *does not try to detect strided streams*. Instead, whenever a core requests for a cache block (e.g., A), the offset prefetcher prefetches the cache block that is distanced by k cache lines (e.g., $A + k$), where k is the *prefetch offset*. In other words, offset prefetchers do not correlate the accessed address to any specific stream; rather, they treat the addresses *individually*, and based on the prefetch offset, they issue a prefetch request for every accessed address. It is noteworthy that the offset prefetcher may adjust the prefetch offset dynamically based on the application's behavior.

In this chapter, we evaluate Spatiotemporal memory streaming (STeMS) [28] and best-offset prefetcher (BOP) [27] as two state-of-the-art nonspatial/temporal prefetching techniques.

2. Spatiotemporal memory streaming (STeMS)

STeMS synergistically integrates spatial and temporal prefetching techniques in a unified prefetcher; STeMS uses a temporal prefetcher to capture the stream of *trigger accesses* (i.e., the first access to each spatial region) and a spatial prefetcher to predict the expected misses *within* the spatial regions. As a spatial prefetcher is incapable of prefetching the *trigger accesses*, these accesses are captured by a temporal prefetcher. The metadata organization of STeMS mainly consists of the metadata tables of STMS [13] and SMS [2].

STeMS, however, seeks to stream the sequence of cache misses *in the order generated by the processor*, regardless of how the corresponding metadata information has been stored in the history tables of STMS and SMS. The reason is that STeMS generates prefetch requests for a large number of addresses. In case it does not prefetch requests in the order generated by the processor, the prefetch requests pollute the cache and hurt performance. To do so, STeMS employs a *Reconstruction Buffer* which is responsible for reordering the prefetch requests generated by the temporal and the spatial prefetchers of STeMS so as to send prefetch requests (and deliver their responses) in the order the processor is supposed to consume them.

For enabling the *reconstruction* process, the metadata tables of SMS and STMS are slightly modified. SMS is modified to record the order of the accessed cache blocks within a spatial region by encoding spatial patterns as ordered lists of offsets, stored in *Patterns Sequence Table (PST)*. Although *PST* is less compact than *PHT* (in the original SMS), the offset lists maintain the order required for accurately interleaving temporal and spatial streams. STMS is also modified and records only spatial triggers (and not all events as in STMS) in a *Region Miss Order Buffer (RMOB)*. Moreover, entries in both spatial and temporal streams are augmented with a *delta* field. The delta field in a spatial (temporal) stream represents the number of events from the temporal (spatial) stream that is interleaved between the current and next events of the same type. Fig. 1 gives an example of how STeMS reconstructs the total miss order.

2.1 Example

Let us assume that the cache block size is 64 bytes. Moreover, the region size for spatial prefetching is 256 bytes.

We consider the following sequence of addresses to see how STeMS behaves.

3, 65, 500, 300, 1000, 900, 129, 220, 400, 500

Of these addresses, 3, 500, 1000 are trigger accesses. We assume that these addresses are generated by instructions PC1, PC2, and PC3, respectively.

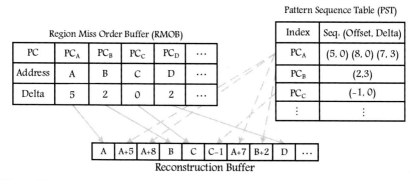

Fig. 1 The organization of spatiotemporal memory streaming (STeMS) and the reconstruction process.

In the beginning, the prefetcher states are empty. Address 3 is the first address to a spatial region, and hence, is a trigger access. Therefore, it will be added to the Region Miss Order Buffer (RMOB) in the form of (address, delta) ==> (0, 0). Note that in RMOB, address is block address, and consequently, the six least significant bits of the address are zero (i.e., block size is 64).

The second memory access is Address 65, which is from the same memory region as the previous access. Therefore, it will be added to the Pattern Sequence Table (PST) in the form of PC: (offset, delta) ==> PC1: (1, 0). Note that offset in PST is from the beginning of the trigger access in terms of cache blocks. The delta is zero because this access and the prior one are from the same spatial region.

The next memory access is Address 500, which is a trigger access. So it will be added to RMOB in the form of (448, 1), where 448 is the block address and 1 is the delta. The delta is one because there is a spatial access between the two trigger accesses.

The next memory access is Address 300, which is from the same memory region as the previous access. Therefore, it will be added to the pattern sequence table (PST) in the form of PC: (offset, delta) ==> PC2: (-3, 0). The delta is zero because this access and the prior one are from the same spatial region.

The next memory access is Address 1000, which is a trigger access. So it will be added to RMOB in the form of (960, 1), where 960 is the block address and 1 is the delta. The delta is one because there is a spatial access between the two trigger accesses.

The next memory access is Address 900, which is from the same memory region as the previous access. Therefore, it will be added to PST in the form of PC: (offset, delta) ==> PC2: (-1, 0).

The next memory access is Address 129, which is from the same memory region as Address 3. Therefore, it will be added to PST in the form of PC: (offset, delta) ==> PC1: (2, 4). The delta is four because there are four memory accesses between Addresses 65 and 129.

The next memory access is Address 220, which is from the same memory region as the previous access. Therefore, it will be added to PST in the form of PC: (offset, delta) ==> PC1: (3, 0).

The next memory access is Address 400, which is from the same memory region as Address 500. Therefore, it will be added to PST in the form of PC: (offset, delta) ==> PC2: (-1, 4). The delta is four because there are four

Table 1 Region miss order buffer (RMOB).

(0, 0)	(448, 1)	(960, 1)

Table 2 Pattern sequence table (PST).

PC1:	(1, 0)	(2, 4)	(3, 0)
PC2:	(-3, 0)	(-1, 4)	
PC3:	(-1, 0)		

memory accesses between Addresses 300 and 400. The content of RMOB and PST after this memory access is shown in Tables 1 and 2, respectively.

The last address is 500, which is a recurring address. STeMS find it in RMOB and prefetches the following sequence of addresses one after another (Note that two addresses are empty).

448,	256,	960,	896,	320

2.2 Best-offset prefetcher (BOP)

BOP is a state-of-the-art proposal for offset prefetching, as well as the winner of the second data prefetching championship (DPC-2) [29]. BOP extends Sandbox Prefetcher (SP) [26], which is the primary proposal for offset prefetching, and enhances its *timeliness*. We first describe the operations of SP and then elucidate how BOP extends it.

SP is an offset prefetcher and attempts to dynamically find the offsets that yield *accurate* prefetch requests. To find such offsets, SP defines an *evaluation period* in which it assesses the prefetching accuracy of multiple predefined offsets, ranging from $-n$ to $+n$, where n is a constant, say, eight. For every prefetch offset, a *score* value is associated, and when the evaluation period is over, only offsets whose score values are beyond a certain threshold are considered accurate offsets; and actual prefetch requests are issued using such offsets.

In the evaluation period, for determining the score values, SP issues *virtual prefetch* requests using various offsets. Virtual prefetching refers to the act of adding the information of candidate prefetch addresses to specific storage rather than actually prefetching them. That is, in the evaluation period, instead of issuing numerous costly prefetch requests using all offsets, prefetch candidates are simply inserted into specific storage. Later, when the application generates actual memory references, the stored prefetch candidates are checked against them: if an actual memory reference matches with a prefetch candidate in the specific storage, it is implied that the candidate was an accurate prefetch request, and accordingly, the score value of the offset that has generated this prefetch candidate is incremented.

For the sake of storage efficiency, SP uses a *Bloom Filter* [30] as the specific storage for keeping the record of prefetch candidates of each offset. Generally, *Bloom Filter* is a probabilistic data structure that is used for examining whether an element is *not* a member of a set. The filter has an array of counters and several hash functions, where each hash function maps the input to a counter in the array. Upon inserting an element into the filter, all counters identified by all of the hash functions are incremented, signifying the membership of the element. Upon checking the membership of an element, all counters identified by all of the hash functions are searched. If at least one of the counters be zero, it is construed that the element is *not* a member of the set as one of the corresponding counters has been incremented. In the context of SP, prefetch candidates generated by offsets are added to the *Bloom Filter*. Then, upon triggering an actual memory reference, *Bloom Filter* is checked to find out if the current memory reference has been inserted into the *Bloom Filter* as a prefetch candidate; and accordingly, the score value of the offsets are manipulated.

Fig. 2 shows the hardware realization of SP that mainly consists of a *Sandbox Prefetch Unit (SPU)* and a *Bloom Filter*. *SPU* maintains the status

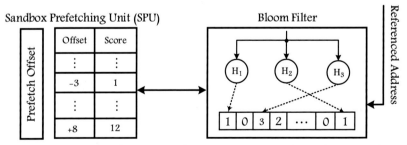

Fig. 2 The organization of Sandbox Prefetcher (SP). The "SPU" keeps the score values of various evaluated offsets. The "Bloom Filter" serves as temporary storage for keeping the prefetch candidates during the evaluation period of each offset. Each *circle* in the "Bloom Filter" represents a different hash function. The "Prefetch Offset" in the figure represents the underevaluation offset.

of several specific offsets and evaluates them in a round-robin fashion. Upon each triggering access (i.e., cache miss or prefetch hit), the cache line address is checked against the *Bloom Filter* to determine if the cache line would have been prefetched by the *underevaluation* offset. If the *Bloom Filter* contains the address, the score value of the offset is incremented; then the procedure repeats for the other offsets. When the evaluation period is over, only offsets whose score values are beyond a certain threshold are allowed to issue prefetch requests.

SP chooses prefetch offsets merely based on their accuracy and ignores the timeliness. Nonetheless, accurate but *late* prefetches do not accelerate the execution of applications as much as timely prefetch requests do. Therefore, BOP tweaks SP and attempts to select offsets that result in timely prefetch requests, having the prefetched blocks ready before the processor actually asks for them.

Fig. 3 shows the hardware structure of BOP. Similar to SP, BOP evaluates the efficiency of various offsets and chooses the best offset for prefetching. However, unlike SP, BOP promotes offsets that generate timely prefetch candidates rather than merely accurate ones. The main idea behind BOP is: "For k to be a timely prefetch offset for line A, line $A - k$ should have been accessed *recently*." That is, offsets whose prefetch candidates are used by the application *not much longer than they generated* are considered as the suitable offsets, and accordingly, their score values are incremented. In order to evaluate the timeliness of prefetch requests

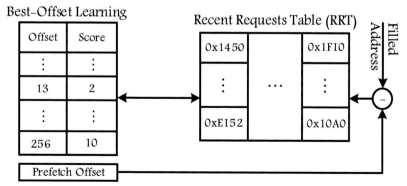

Fig. 3 The organization of Best-Offset Prefetcher (BOP). The "Best-Offset Learning" keeps track of score values associated with various offsets, serving the function of "SPU" in SP. The "Recent Requests Table" holds the recent prefetch candidates. The "Prefetch Offset" in the figure represents the underevaluation offset.

issued using various offsets, BOP replaces SP's *Bloom Filter* with a set-associative *Recent Requests Table (RRT)*. The size of the *RRT* is purposely chosen to be small in order to keep *only* recent requests. For every triggering event A, under the evaluation period of offset k, the score of the offset is incremented if line $A - k$ hits in the *RRT*. In other words, under the evaluation period of offset k, if line A is requested by the processor and line $A - k$ hits in the *RRT*, it is construed that "if offset k had issued a prefetch request upon line $A - k$, its prefetch requests (i.e., $A - k + k = A$) would have been timely." Therefore, k is identified as an offset whose prefetch requests match with the demand of the processor, and thus, its score is incremented.

Unlike SP that evaluates offsets in the range of $- n$ to $+ n$ for some constant n, BOP evaluates 46 (almost random) constant offsets that were picked empirically. The other components and functions of BOP are similar to those of SP.

3. Summary

Spatiotemporal prefetching has been proposed and developed to capture both temporal and spatial memory access patterns of applications. Spatiotemporal prefetching is based on the observation that temporal and spatial prefetchers each target a specific subset of cache misses, and leave the rest uncovered. Spatialtemporal data prefetching tries to synergistically capture both types of patterns, that any of the temporal or spatial prefetcher lonely cannot. Offset prefetching considers low overhead as a major design constraint in architecting the prefetcher. With offset prefetching, whenever a core request a cache block, the offset prefetcher prefetches the cache block that is distanced by a certain offset. Offset prefetching has a very simple structure and has been show quite effective for a variety of applications.

References

[1] M. Bakhshalipour, M. Shakerinava, P. Lotfi-Kamran, H. Sarbazi-Azad, Bingo spatial data prefetcher, in: Proceedings of the International Symposium on High-Performance Computer Architecture (HPCA), 2019, pp. 399–411.

[2] S. Somogyi, T.F. Wenisch, A. Ailamaki, B. Falsafi, A. Moshovos, Spatial memory streaming, in: Proceedings of the International Symposium on Computer Architecture (ISCA), 2006, pp. 252–263.

[3] K.J. Nesbit, J.E. Smith, Data cache prefetching using a global history buffer, in: Proceedings of the International Symposium on High Performance Computer Architecture (HPCA), 2004, p. 96.

[4] M. Shevgoor, S. Koladiya, R. Balasubramonian, C. Wilkerson, S.H. Pugsley, Z. Chishti, Efficiently prefetching complex address patterns, in: Proceedings of the International Symposium on Microarchitecture (MICRO), 2015, pp. 141–152.

[5] K.J. Nesbit, A.S. Dhodapkar, J.E. Smith, AC/DC: an adaptive data cache prefetcher, in: Proceedings of the International Conference on Parallel Architectures and Compilation Techniques (PACT), 2004, pp. 135–145.

[6] S. Kumar, C. Wilkerson, Exploiting spatial locality in data caches using spatial footprints, in: Proceedings of the International Symposium on Computer Architecture (ISCA), 1998, pp. 357–368.

[7] C.F. Chen, S.-H. Yang, B. Falsafi, A. Moshovos, Accurate and complexity-effective spatial pattern prediction, in: Proceedings of the International Symposium on High Performance Computer Architecture (HPCA), 2004, pp. 276–287.

[8] J.F. Cantin, M.H. Lipasti, J.E. Smith, Stealth prefetching, in: Proceedings of the International Conference on Architectural Support for Programming Languages and Operating Systems (ASPLOS), 2006, pp. 274–282.

[9] J. Kim, S.H. Pugsley, P.V. Gratz, A.L.N. Reddy, C. Wilkerson, Z. Chishti, Path confidence based lookahead prefetching, in: Proceedings of the International Symposium on Microarchitecture (MICRO), 2016, pp. 60:1–60:12.

[10] M. Bakhshalipour, M. Shakerinava, P. Lotfi-Kamran, H. Sarbazi-Azad, Accurately and maximally prefetching spatial data access patterns with bingo, in: The Third Data Prefetching Championship, 2019.

[11] D. Joseph, D. Grunwald, Prefetching using Markov predictors, in: Proceedings of the International Symposium on Computer Architecture (ISCA), 1997, pp. 252–263.

[12] Y. Chou, Low-cost epoch-based correlation prefetching for commercial applications, in: Proceedings of the International Symposium on Microarchitecture (MICRO), 2007, pp. 301–313.

[13] T.F. Wenisch, M. Ferdman, A. Ailamaki, B. Falsafi, A. Moshovos, Practical off-chip meta-data for temporal memory streaming, in: Proceedings of the International Symposium on High Performance Computer Architecture (HPCA), 2009, pp. 79–90.

[14] T.F. Wenisch, S. Somogyi, N. Hardavellas, J. Kim, A. Ailamaki, B. Falsafi, Temporal streaming of shared memory, in: Proceedings of the International Symposium on Computer Architecture (ISCA), 2005, pp. 222–233.

[15] Y. Solihin, J. Lee, J. Torrellas, Using a user-level memory thread for correlation prefetching, in: Proceedings of the International Symposium on Computer Architecture (ISCA), 2002, pp. 171–182.

[16] A. Jain, C. Lin, Linearizing irregular memory accesses for improved correlated prefetching, in: Proceedings of the International Symposium on Microarchitecture (MICRO), 2013, pp. 247–259.

[17] M. Bakhshalipour, P. Lotfi-Kamran, H. Sarbazi-Azad, Domino temporal data prefetcher, in: Proceedings of the International Symposium on High-Performance Computer Architecture (HPCA), IEEE, 2018, pp. 131–142.

[18] M. Bakhshalipour, P. Lotfi-Kamran, H. Sarbazi-Azad, An efficient temporal data prefetcher for L1 caches, IEEE Comput. Architec. Lett. CAL 16 (2) (2017) 99–102.

[19] M. Bakhshalipour, P. Lotfi-Kamran, A. Mazloumi, F. Samandi, M. Naderan-Tahan, M. Modarressi, H. Sarbazi-Azad, Fast data delivery for many-core processors, IEEE Trans. Comput. (TC) 67 (10) (2018) 1416–1429.

[20] E. Ebrahimi, O. Mutlu, Y.N. Patt, Techniques for bandwidth-efficient prefetching of linked data structures in hybrid prefetching systems, in: Proceedings of the International Symposium on High-Performance Computer Architecture (HPCA), 2009, pp. 7–17.

[21] M. Ferdman, B. Falsafi, Last-touch correlated data streaming, in: Proceedings of the International Symposium on Performance Analysis of Systems & Software (ISPASS), 2007, pp. 105–115.

[22] C.J. Hughes, S.V. Adve, Memory-side prefetching for linked data structures for processor-in-memory systems, J. Parallel Distrib. Comput. 65 (4) (2005) 448–463.

[23] A.-C. Lai, C. Fide, B. Falsafi, Dead-block prediction & dead-block correlating prefetchers, in: Proceedings of the International Symposium on Computer Architecture (ISCA), 2001, pp. 144–154.

[24] F. Golshan, M. Bakhshalipour, M. Shakerinava, A. Ansari, P. Lotfi-Kamran, H. Sarbazi-Azad, Harnessing pairwise-correlating data prefetching with runahead metadata, IEEE Comput. Architec. Lett. (CAL) 19 (2) (2020) 130–133, https://doi.org/10.1109/LCA.2020.3019343.

[25] A. Roth, G.S. Sohi, Effective jump-pointer prefetching for linked data structures, in: Proceedings of the International Symposium on Computer Architecture (ISCA), 1999, pp. 111–121.

[26] S.H. Pugsley, Z. Chishti, C. Wilkerson, P.-f. Chuang, R.L. Scott, A. Jaleel, S.-L. Lu, K. Chow, R. Balasubramonian, Sandbox prefetching: safe run-time evaluation of aggressive prefetchers, in: Proceedings of the International Symposium on High-Performance Computer Architecture (HPCA), 2014, pp. 626–637.

[27] P. Michaud, Best-offset hardware prefetching, in: Proceedings of the International Symposium on High-Performance Computer Architecture (HPCA), 2016, pp. 469–480.

[28] S. Somogyi, T.F. Wenisch, A. Ailamaki, B. Falsafi, Spatio-temporal memory streaming, in: Proceedings of the International Symposium on Computer Architecture (ISCA), 2009, pp. 69–80.

[29] S. Pugsley, A. Alameldeen, C. Wilkerson, H. Kim, The second data prefetching championship (DPC-2), IEEE, 2015.

[30] B.H. Bloom, Space/time trade-offs in hash coding with allowable errors, Commun. ACM 13 (7) (1970) 422–426.

About the authors

Pejman Lotfi-Kamran is an associate professor and the head of the school of computer science and the director of Turin Cloud Services at Institute for Research in Fundamental Sciences (IPM). His research interests include computer architecture, computer systems, approximate computing, and cloud computing. His work on scale-out server processor design lays the foundation for Cavium ThunderX. Lotfi-Kamran has a Ph.D. in computer science from the École Polytechnique Fédérale de Lausanne (EPFL). He received his MS and BS in computer engineering from the University of Tehran. He is a member of the IEEE and the ACM.

 **Hamid Sarbazi-Azad** is currently a professor of computer science and engineering at the Sharif University of Technology, Tehran, Iran. His research interests include high-performance computer/memory architectures, NoCs and SoCs, parallel and distributed systems, social networks, and storage systems, on which he has published over 400 refereed papers. He received Khwarizmi International Award in 2006, TWAS Young Scientist Award in engineering sciences in 2007, Sharif University Distinguished Researcher awards in the years 2004, 2007, 2008, 2010, and 2013, the Iranian Ministry of Communication and Information Technology's award for contribution to IT research and education in 2014, and distinguished book author of the Sharif University of Technology in 2015 and 2020. Dr Sarbazi-Azad is now an associate editor of ACM Computing Surveys, IEEE Computer Architecture Letters, and Elsevier's Computers and Electrical Engineering.

CHAPTER FIVE

State-of-the-art data prefetchers

Mehran Shakerinava[a], Fatemeh Golshan[a], Ali Ansari[a], Pejman Lotfi-Kamran[b], and Hamid Sarbazi-Azad[c]

[a]Sharif University of Technology, Tehran, Iran
[b]School of Computer Science, Institute for Research in Fundamental Sciences (IPM), Tehran, Iran
[c]Sharif University of Technology, and Institute for Research in Fundamental Sciences (IPM), Tehran, Iran

Contents

Abstract

We introduced several styles of data prefetching in the past three chapters. The introduced data prefetchers were known for a long time, sometimes for decades. In this chapter, we introduce several state-of-the-art data prefetchers, which have been introduced in the past few years. In particular, we introduce Domino, Bingo, MLOP, and Runahead Metadata.

In this chapter, we describe our own proposals for efficient data prefetching that have been published in recent years. We include them in chronological order based on their publication date.

1. Domino temporal data prefetcher

Domino is a state-of-the-art temporal data prefetcher that is built upon STMS and seeks to improve its effectiveness. Domino is based on the observation that a single miss address, as used in the lookup mechanism of STMS, cannot always identify the correct miss stream in the history. Therefore, Domino provides a mechanism to look up the history of miss addresses with a combination of the last *one or two* miss addresses. To do so, Domino replaces the *Index Table* of STMS with a novel structure, named *Enhanced*

Index Table (EIT). *EIT* like the *Index Table* of STMS stores a pointer for each address in the history; but unlike it, *keeps the subsequent miss of each address,* additionally. Having the next miss of every address in the *EIT* enables DOMINO to find the correct stream in the history using the last one or two misses addresses. Moreover, with this organization, DOMINO becomes able to start prefetching (i.e., issuing the first prefetch request) right after touching *EIT*. That is, unlike STMS that needs to wait for two serial memory accesses (one for *Index Table,* then another for *History Table)* to start prefetching, DOMINO can start prefetching immediately after accessing *EIT,* because *EIT* contains the address of the first prefetch candidate. Starting prefetching sooner, causes DOMINO to offer superior timeliness as compared to STMS. Fig. 1 shows the organization of the *EIT*.

The *EIT* is indexed by a *single* miss address. Associated with every tag, there are several address–pointer pairs, where the address is a miss of the core and the pointer is a location in the *History Table*. An *(a, p)* pair associated to tag *t* indicates that the pointer to the last occurrence of miss address *t* followed by *a* is *p*. The tag along with its associated address–pointer pairs is called a *super-entry,* and every address–pointer pair is named an *entry*. Every *row* of the *EIT* has several *super-entries,* and each *super-entry* has several *entries.* DOMINO keeps the LRU stack among both the *super-entries* and the *entries* within each *super-entry*. Upon a cache miss, DOMINO uses the missed address to fetch a *row* of the *EIT*. Then, DOMINO attempts to find the *super-entry* associated with the missed address. In case a match is not found, nothing will be done, and otherwise, a prefetch will be sent for the address field of the most recent *entry* in the found *super-entry*. When the next triggering event occurs (miss or prefetch hit), DOMINO searches the *super-entry* and picks the *entry* for which the address field matches the triggering event. In case no match is found, DOMINO uses the triggering event to bring another *row* from the

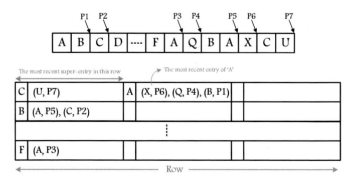

Fig. 1 The details of the Enhanced Index Table in Domino prefetcher [1].

EIT. Otherwise, Domino sends a request to read the *row* of the *History Table* pointed to by the pointer field of the matched *entry.* Once the sequence of miss addresses from the *History Table* arrives, Domino issues prefetch requests.

2. Bingo spatial data prefetcher

Bingo is a recent proposal for spatial data prefetching, as well as the runner-up (and the winner in the multi-core evaluations) of The Third Data Prefetching Championship (DPC-3) [2]. Bingo is based on the observation that assigning footprint information to a single event is suboptimal as compared to a case where footprints are correlated with *multiple* events. Bingo discusses that events either are *short,* that while their probability of recurrence is high, assigning footprints to them results in low accuracy, or are *long,* that while prefetching using them results in high accuracy, much of the opportunity gets lost since the probability of recurring them is quite low. Bingo, for this reason, proposes to associate the observed footprint information to multiple events in order to provide both high opportunity and high accuracy. More specifically, Bingo correlates the observed footprint information of various pages with both *PC + Address* and *PC + Offset* of trigger accesses. In the context of Bingo, *PC + Address* is considered as a long event, while *PC + Offset* is treated as a short event. Whenever the time for prefetching comes (i.e., a triggering access occurs), Bingo uses the footprint that is associated with the longest occurred event for prefetching (i.e., *PC + Address;* and, if no history is recorded for *PC + Address, PC + Offset*).

A naive implementation of Bingo requires two distinct *PHTs:* one table maintains the history of footprints observed after each *PC + Address,* while the other keeps the footprint metadata associated with *PC + Offset.* Upon looking for a pattern to prefetch, logically, first, the *PC + Address* of the trigger access is used to search the long *PHT.* If a match is found, the corresponding footprint is utilized to issue prefetch requests. Otherwise, the *PC + Offset* of the trigger access is used to look up the short *PHT.* In case of a match, the footprint metadata of the matched entry will be used for prefetching. If no matching entry is found, no prefetch will be issued. Such an implementation, however, would impose significant storage overhead. Authors in Bingo observe that, in the context of spatial data prefetching, a considerable fraction of the metadata that is stored in the *PHTs* are *redundant.* That is, there are many cases where both metadata tables (tables associated with long and short events) offer the same prediction.

To efficiently eliminate redundancies in the metadata storage, instead of using multiple history tables, BINGO proposes *having a single history table but looking it up multiple times, each time with a different event.* Fig. 2 details the practical design of BINGO which uses only one *PHT.* The main insight is that *short events are carried in long events.* That is, by having the long event at hand, one can find out what the short events are, just by ignoring parts of the long event. For the case of BINGO, the information of *PC + Offset* is carried in *PC + Address;* therefore, by knowing the *PC + Address,* the *PC + Offset* is also known. To exploit this phenomenon, BINGO proposes having only one history table which *stores just the history of the long event but is looked up with both long and short events.* For the case of BINGO, the history table stores footprints which were observed after each *PC + Address* event, but is looked up with both the *PC + Address* and *PC + Offset* of the trigger access in order to offer high accuracy while not losing prefetching opportunities.

To enable this, BINGO *indexes the table with a hash of the shortest event* but *tags it with the longest event.* Whenever a piece of information is going to be stored in the history metadata, it is associated with the longest event, and then stored in the history table. To do so, *the bits corresponding to the shortest*

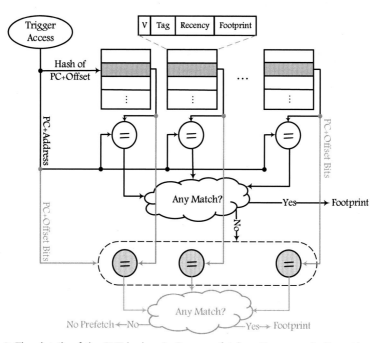

Fig. 2 The details of the PHT lookup in BINGO prefetcher. Gray parts indicate the case where lookup with long event fails to find a match. Each large rectangle indicates a physical way of the history table.

event are used for indexing the history table to find the set in which the metadata should be stored; however, *all bits of the longest event are used to tag the entry.* More specifically, with BINGO, whenever a new footprint is going to be stored in the metadata organization, it is associated with the corresponding *PC + Address.* To find a location in the history table for the new entry, a hash of only *PC + Offset* is used to index the table. By knowing the set, the baseline replacement algorithm (e.g., LRU) is used to choose a victim to open room for storing the new entry. After determining the location, the entry is stored in the history table, but all bits of the *PC + Address* are used for tagging the entry.

Whenever there is a need for prediction, the *PHT* is first looked up with the longest event; if a match is found, it will be used to make a prediction. Otherwise, the table should be looked up with the next-longest event. As both long and short events are mapped to the same set, there is no need to check a new set; instead, the entries of the same set are tested to find a match with the shorter event. With BINGO, the table is first looked up with the *PC + Address* of the trigger access. If a match is found, the corresponding footprint metadata will be used for issuing prefetch requests. Otherwise, the table should be looked up with the *PC + Offset* of the trigger access. As both *PC + Address* and *PC + Offset* are mapped to the same set, there is no need to check a new set. That is, all the corresponding *PC + Offset* entries *should be in the same set.* Therefore, the entries of the same set are tested to find a match. In this case, however, not all bits of the stored tags in the entries are necessary to match; only the *PC + Offset* bits need to be matched. This way, BINGO associates each footprint with more than one event (i.e., both *PC + Address* and *PC + Offset*) but store the footprint metadata in the table with only one of them (the longer one) to reduce the storage requirement. Doing so, redundancies are automatically eliminated because a metadata footprint is stored once with its *PC + Address* tag.

3. Multi-lookahead offset prefetcher

MULTI-LOOKAHEAD OFFSET PREFETCHER (MLOP) [3] is a state-of-the-art *offset prefetching*. Offset prefetching, in fact, is an evolution of stride prefetching, in which, the prefetcher does *not* try to detect strided streams. Instead, whenever a core requests for a cache block (e.g., A), the offset prefetcher prefetches the cache block that is distanced by k cache lines (e.g., $A + k$), where k is the *prefetch offset*. In other words, offset prefetchers do not correlate the accessed address to any specific stream; rather, they treat the addresses individually, and based on the prefetch offset, they issue a prefetch request for every accessed address. Offset prefetchers have been shown to offer significant performance benefits while imposing small storage and logic overheads [4,5].

The initial proposal for offset prefetching, named SANDBOX PREFETCHER (SP) [4], attempts to find offsets that yield *accurate* prefetch requests. To find such offsets, SP evaluates the prefetching accuracy of several predefined offsets (e.g., −8, −7, …, +8) and finally allows offsets whose prefetching accuracy are beyond a certain threshold to issue actual prefetch requests. The later work, named BEST-OFFSET PREFETCHER (BOP) [5] tweaks SP and sets the *timeliness* as the evaluation metric. BOP is based on the insight that accurate but *late* prefetch requests do not accelerate the execution of applications as much as timely requests do. Therefore, BOP finds offsets that yield timely prefetch requests in an attempt to have the prefetched blocks ready before the processor actually asks for them.

MLOP takes another step and proposes a novel offset prefetcher. MLOP is based on the observation that while BOP is able to generate timely prefetch requests, it loses much opportunity at covering cache misses because of *relying on a single best offset and discarding many other appropriate offsets*. BOP evaluates several offsets and considers the offset that can generate the most timely prefetch requests as the best offset; then, it relies *only* on this best offset to issue prefetch requests until another offset becomes better, and hence, the new best. In fact, this is a binary classification: the prefetch offsets are considered either as *timely* offsets or *late* offsets. After classification, the prefetcher does *not* allow the so-called late offsets to issue any prefetch requests. However, there might be many other appropriate offsets that are less timely but are of value in that they can hide a significant fraction of cache miss delays.

To overcome the deficiencies of prior work, MLOP proposes to have a *spectrum of timelinesses* for various prefetch offsets during their evaluations, rather than binarily classifying them. During the evaluation of various prefetch offsets, MLOP considers *multiple lookaheads* for every prefetch offset: *with which lookahead can an offset cover a specific cache miss?* To implement this, MLOP considers several lookaheads for each offset, and assigns score values to every offset with every lookahead, *individually*. Finally, when the time for prefetching comes, MLOP finds the *best offset for each lookahead* and allows it to issue prefetch requests; however, the prefetch requests for smaller lookaheads are *prioritized* and issued first. By doing so, MLOP ensures that it allows the prefetcher to issue enough prefetch requests (i.e., various prefetch offsets are utilized; high miss coverage) while the timeliness is well considered (i.e., the prefetch requests are ordered).

Fig. 3 shows an overview of MLOP. To extract offsets from access patterns, MLOP uses an *Access Map Table (AMT)*. The *AMT* keeps track of several recently-accessed addresses, along with a bit-vector for each base address. Each bit in the bit-vector corresponds to a cache block in the neighborhood of the address, indicating whether or not the block has been accessed.

MLOP considers an evaluation period in which it evaluates several prefetch offsets and chooses the qualified ones for issuing prefetch requests later on. For every offset, it considers multiple levels of score where each level corresponds to a specific lookahead. The score of an offset at lookahead level X indicates the number of cases where the offset prefetcher could have prefetched an access, at least X accesses prior to occurrence. For example, the score of offsets at the lookahead level 1 indicates the number of cases where the offset prefetcher could have prefetched any of the futures accesses.

To efficiently mitigate the negative effect of all predictable cache misses, MLOP selects *one best offset from each lookahead level*. Then, during the actual prefetching, it allows all selected best offsets to issue prefetch requests. Doing so, MLOP ensures that it chooses enough prefetch offsets (i.e., does not suppress many qualified offsets like prior work [5]), and will cover a significant fraction of cache misses that are predictable by offset prefetching. To handle the timeliness issue, MLOP ties to send the prefetch requests in a way that the application would have sent if there had not been any prefetcher: MLOP starts from lookahead level 1 (i.e., the accesses that are expected to happen the soonest) and issues the corresponding prefetch requests (using its best offset), then goes to the upper level; this process repeats. With this prioritization, MLOP tries to hide the latency of all predictable cache misses, as much as possible.

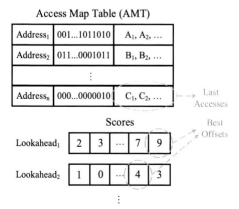

Fig. 3 The hardware realization of our MLOP.

4. Runahead metadata

RUNAHEAD METADATA (RMD) [6] is a general technique for harnessing *pairwise-correlating* prefetching. Pairwise-correlating prefetching refers to methods that correlate every address (or more generally, every *event* [7]) with a *single next prediction*. The next prediction can be the next expected address, with address-based pairwise-correlating prefetchers [8,9,10,11,12] or the next expected delta, with delta-based pairwise-correlating prefetchers [13,14,5].

A main challenge in pairwise-correlating prefetching is harnessing prefetching degree. Unlike streaming prefetchers [15,16,1] that prefetch multiple data addresses that follow the correlated address in the FIFO history buffer, or footprint-based prefetchers [17,18,7] that prefetch multiple data addresses whose corresponding bit in the bit-vector is set, pairwise-correlating prefetchers are limited to a single prediction per correlation entry; they cannot trivially issue *multi-degree* prefetching. With this lookahead limitation, pairwise-correlating prefetching faces *timeliness* as a major problem, in that, issuing merely a single prefetch request every time may not result in prefetch requests that cover the *whole* latency of cache misses.

What is typically employed in state-of-the-art pairwise-correlating data prefetchers as the de facto mechanism, including delta-based [14] and address-based [12] ones, and even similar instruction prefetchers [19], is *using the prediction as input to the metadata tables to make more predictions:* whenever a prediction is made, the prefetcher assumes it a correct prediction, and repeatedly indexes the metadata table with the prediction to make more predictions. While this approach has no storage overhead, it offers poor accuracy, as explicitly shown by recent work [13,7,1,19]. The problem with such an approach is that the prefetcher has no information about *how many times it should repeat this process*. In fact, this emanates from dissimilar stream lengths: if the prefetcher repeats this process N times, for streams whose length is smaller than N, say M, it overprefetches $N - M$ addresses, resulting in inaccuracy; for streams longer than N, it may lose timeliness. Prior approaches that perform multi-degree prefetching in such a way, typically choose the degree of prefetching empirically based on a set of studied workloads. For example, Shevgoor et al. [14] set the degree to four; Bakhshalipour et al. [12] set it to three. These numbers are chosen completely experimentally for a specific configuration and by examining a limited number of workloads, with which, the chosen number provides a reasonable trade-off between accuracy and timeliness. Obviously, limiting the degree to a certain predefined number neither is a solution that scales to various configurations and workloads, nor is optimal (w.r.t accuracy and timeliness) for the examined very configuration/workloads.

RMD proposes a novel solution to harness the multi-degree prefetching in the context of pairwise-correlating prefetchers. The key idea is to *have separate metadata information for predicting the next but one expected event (e.g., the delta following the next delta; two deltas away from now)*. This way, in fact, RMD employs two separate metadata tables: one predicts the next event (*Distance1*; *D1*), the other predicts the next but one event *(Distance2*; *D2*), which is called *Runahead Metadata Table*. When issuing multi-degree prefetching, the first prefetching is issued using only *D1*. For issuing the second prefetch, *D1* is searched using its previous prediction, similar to multi-degree prefetching of previous methods; meanwhile, *D2* is searched using the actual input (not prediction); the prefetch request is issued only if the prediction of both tables match; otherwise, the prefetching is finished. From the third prefetch request (if any) onward, both tables are searched using the corresponding inputs from the previous steps; if their predictions match, the prefetch request is issued and the process continues; otherwise, the prefetching is finished, concluding that the stream has come to an end.

The reason for adding *D2* is to *harness* the multi-degree prefetching of *D1*: until when the recursive lookups should resume? As *D2* operates one step ahead of *D1*, what *D2* offers is what *D1* is expected to offer in the next step. Hence, when *D1*'s second prediction (i.e., prediction using the previous prediction as input) is *not* equal with *D2*'s prediction, it is intuitively concluded that the stream has been finished, and no further prefetch request is issued for the current stream. However, as long as the predictions match, the prefetcher continues prefetching to provide efficient timeliness, while preserving accuracy.

Fig. 4 epitomizes how *RMD* works. The entries in tables are interpreted in this way: $<A, B>$ in *D1* shows that immediately after *A*, it is expected *B* to happen; $<C, J>$ in *D2* is intended to mean that two steps away from *C*, it is expected *J* to happen. Consider that *A* happens. RMD indexes *D1* by *A*. The prediction of *D1* is *B*; a prefetch request is issued for *B*. Then, *D1* is indexed by *B*; meanwhile, *D2* is indexed by *A*. The predictions of both *D1* and *D2* are *C*; their predictions match, and the prefetcher prefetches *C*. Then, *D1* is indexed by *C* and *D2* by *B*. The prediction of *D1* is *D*, and the prediction of *D2* is *P*; their predictions do not match, and no further prefetch request is issued.

D1 Table

Delta	Pred.
A	B
B	C
C	D

D2 Table

Delta	Pred.
A	C
B	P
C	J

Fig. 4 An illustration of how RMD works.

5. Summary

Prefetching is an active area of research. After four decades of research, there are still significant challenges/opportunities in this area. In our recent research, we address maximal and accurate extraction of patterns in applications with (extremely) irregular memory accesses, metadata overhead reduction, and controlling the aggressiveness of prefetching. We envisage future work will further improve the effectiveness of prefetching in the presence of irregular applications, improving upon our work. More, we expect that future work target further challenges/opportunities in prefetching including but not limited to modern applications (e.g., deep learning), heterogeneous computing platforms (e.g., CPU + GPU), heterogenous memory systems (e.g., DRAM + NVM), and further.

References

[1] M. Bakhshalipour, P. Lotfi-Kamran, H. Sarbazi-Azad, Domino temporal data prefetcher, in: Proceedings of the International Symposium on High-Performance Computer Architecture (HPCA), IEEE, 2018, pp. 131–142.

[2] The Third Data Prefetching Championship, 2019. https://dpc3.compas.cs.stonybrook.edu/.

[3] M. Shakerinava, M. Bakhshalipour, P. Lotfi-Kamran, H. Sarbazi-Azad, Multi-lookahead offset prefetching, in: The Third Data Prefetching Championship, 2019.

[4] S. H. Pugsley, Z. Chishti, C. Wilkerson, P.-f. Chuang, R. L. Scott, A. Jaleel, S.-L. Lu, K. Chow, and R. Balasubramanian, "Sandbox prefetching: safe run-time evaluation of aggressive Prefetchers," in Proceedings of the International Symposium on High-Performance Computer Architecture (HPCA), 2014, pp. 626–637.

[5] P. Michaud, Best-offset hardware prefetching, in: Proceedings of the International Symposium on High-Performance Computer Architecture (HPCA), 2016, pp. 469–480.

[6] F. Golshan, M. Bakhshalipour, M. Shakerinava, A. Ansari, P. Lotfi-Kamran, H. Sarbazi-Azad, Harnessing pairwise-correlating data prefetching with runahead metadata, IEEE Computer Architecture Letters (CAL) (2020).

[7] M. Bakhshalipour, M. Shakerinava, P. Lotfi-Kamran, H. Sarbazi-Azad, Bingo spatial data prefetcher, in: Proceedings of the International Symposium on High-Performance Computer Architecture (HPCA), 2019, pp. 399–411.

[8] D. Joseph, D. Grunwald, Prefetching using markov predictors, in: Proceedings of the International Symposium on Computer Architecture (ISCA), 1997, pp. 252–263.

[9] Z. Hu, S. Kaxiras, M. Martonosi, Timekeeping in the memory system: predicting and optimizing memory behavior, in: Proceedings of the International Symposium on Computer Architecture (ISCA), 2002, pp. 209–220.

[10] Z. Hu, M. Martonosi, S. Kaxiras, TCP: tag correlating prefetchers, in: Proceedings of the International Symposium on High Performance Computer Architecture (HPCA), 2003, pp. 317–326.

[11] A.-C. Lai, C. Fide, B. Falsafi, Dead-block prediction & dead-block correlating prefetchers, in: Proceedings of the International Symposium on Computer Architecture (ISCA), 2001, pp. 144–154.

[12] M. Bakhshalipour, P. Lotfi-Kamran, H. Sarbazi-Azad, An efficient temporal data prefetcher for L1 caches, IEEE Computer Architecture Letters (CAL) 16 (2) (2017) 99–102.

[13] J. Kim, S.H. Pugsley, P.V. Gratz, A.L.N. Reddy, C. Wilkerson, Z. Chishti, Path confidence based lookahead prefetching, in: Proceedings of the International Symposium on Microarchitecture (MICRO), 2016. pp. 60:1–60:12.

[14] M. Shevgoor, S. Koladiya, R. Balasubramonian, C. Wilkerson, S.H. Pugsley, Z. Chishti, Efficiently prefetching complex address patterns, in: Proceedings of the International Symposium on Microarchitecture (MICRO), 2015, pp. 141–152.

[15] T.F. Wenisch, S. Somogyi, N. Hardavellas, J. Kim, A. Ailamaki, B. Falsafi, Temporal streaming of shared memory, in: Proceedings of the International Symposium on Computer Architecture (ISCA), 2005, pp. 222–233.

[16] T.F. Wenisch, M. Ferdman, A. Ailamaki, B. Falsafi, A. Moshovos, Practical off-chip meta-data for temporal memory streaming, in: Proceedings of the International Symposium on High Performance Computer Architecture (HPCA), 2009, pp. 79–90.

[17] S. Kumar, C. Wilkerson, Exploiting spatial locality in data caches using spatial footprints, in: Proceedings of the International Symposium on Computer Architecture (ISCA), 1998, pp. 357–368.

[18] S. Somogyi, T.F. Wenisch, A. Ailamaki, B. Falsafi, A. Moshovos, Spatial memory streaming, in: Proceedings of the International Symposium on Computer Architecture (ISCA), 2006, pp. 252–263.

[19] L. Spracklen, Y. Chou, S.G. Abraham, Effective instruction prefetching in chip multiprocessors for modern commercial applications, in: Proceedings of the International Symposium on High-Performance Computer Architecture (HPCA), 2005, pp. 225–236.

About the authors

Mehran Shakerinava received his B.Sc. in Computer Engineering from Sharif University of Technology, Iran (2019). He is currently pursuing a Ph.D. in Computer Science at McGill University and Mila Quebec AI Institute.

Fatemeh Golshan received the B.Sc. degree in Computer Engineering from Isfahan University of Technology, Isfahan, Iran, in 2018, and the M.Sc. degree in Computer Architecture from the Sharif University of Technology, Tehran, Iran, in 2020. Her research interest is Computer Architecture with an emphasis on memory systems.

Ali Ansari is a Ph.D student at the École Polytechnique Fédérale de Lausanne (EPFL) pursuing his program under the supervision of professor Babak Falsafi. He graduated from the master's program from the Sharif University of Technology in Tehran, Iran, in 2019. During his master studies, he researched instruction and data prefetching techniques advised by professors Hamid Sarbazi-Azad and Pejman Lotfi-Kamran. His research interests span high-performance computer architecture, memory systems, and emerging workloads.

Pejman Lotfi-Kamran is an associate professor and the head of the School of Computer Science and the director of Turin Cloud Services at Institute for Research in Fundamental Sciences (IPM). His research interests include computer architecture, computer systems, approximate computing, and cloud computing. His work on scale-out server processor design lays the foundation for Cavium ThunderX. Lotfi-Kamran has a Ph.D. in Computer Science from the École Polytechnique Fédérale de Lausanne (EPFL). He received his MS and BS in Computer Engineering from the University of Tehran. He is a member of the IEEE and the ACM.

Hamid Sarbazi-Azad is currently a professor of Computer Science and Engineering at the Sharif University of Technology, Tehran, Iran. His research interests include high-performance computer/memory architectures, NoCs and SoCs, parallel and distributed systems, social networks, and storage systems, on which he has published over 400 refereed papers. He received Khwarizmi International Award in 2006, TWAS Young Scientist Award in Engineering Sciences in 2007, Sharif University Distinguished Researcher awards in the years 2004, 2007, 2008, 2010, and 2013, the Iranian Ministry of Communication and Information Technology's award for contribution to IT research and education in 2014, and distinguished book author of the Sharif University of Technology in 2015 and 2020. Dr Sarbazi-Azad is now an associate editor of ACM Computing Surveys, IEEE Computer Architecture Letters, and Elsevier's Computers and Electrical Engineering.

CHAPTER SIX

Evaluation of data prefetchers

Mehran Shakerinava[a], Fatemeh Golshan[a], Ali Ansari[a], Pejman Lotfi-Kamran[b], and Hamid Sarbazi-Azad[c]
[a]Sharif University of Technology, Tehran, Iran
[b]School of Computer Science, Institute for Research in Fundamental Sciences (IPM), Tehran, Iran
[c]Sharif University of Technology, and Institute for Research in Fundamental Sciences (IPM), Tehran, Iran

Contents

Abstract

We introduced several data prefetchers and qualitatively discussed their strengths and weaknesses. Without quantitative evaluation, the true strengths and weaknesses of a data prefetcher are still vague. To shed light on the strengths and weaknesses of the introduced data prefetchers and to enable the readers to better understand these prefetchers, in this chapter, we quantitatively compare and contrast them.

1. Introduction

In this chapter, we evaluate the effectiveness of data prefetchers in three aspects: miss coverage (the percentage of cache misses eliminated by the prefetcher), overprediction (the erroneous prefetchers normalized to the number of cache misses without prefetching), and application speedup. We evaluate various classes of data prefetchers separately: spatial prefetching, temporal Prefetching, spatio-temporal prefetching, offset prefetching. In addition, we also include a separate section to evaluate multi-degree prefetching with pair-wise correlating methods. In each section, we first explain the methodology of the evaluation and then proceed with the results.

2. Spatial prefetching

2.1 Methodology

We evaluate spatial prefetchers, SMS [1], VLDP [2], and Bingo [3]. We use ChampSim[a] [4], the simulation infrastructure used in the Second Data Prefetching Championship (DPC-2) [5], to meticulously simulate a system whose configuration is shown in Table 1. We model a system based on Intel's recent Xeon Processor [6]. The chip has four OoO cores with an 8 MB shared last-level cache (LLC). Two memory channels are used for accessing off-chip DRAM, providing a maximum bandwidth of

Table 1 Spatial prefetching: Evaluation parameters.

Parameter	Value
Chip	14 nm, 4 GHz, 4 cores
Cores	4-wide OoO, 256-entry ROB, 64-entry LSQ
Fetch Unit	Perceptron [9], 16-entry pre-dispatch queue
L1-D/I	Split I/D, 64 KB, 8-way, 8-entry MSHR
L2 Cache	8 MB, 16-way, 4 banks, 15-cycle hit latency
Main Memory	60 ns zero-load latency, 37.5 GB/s peak bandwidth

---

[a] The source code of our simulator and the implementation of evaluated data prefetchers are available at https://github.com/bakhshalipour/Bingo.

37.5 GB/s. The OS uses 4 KB pages and virtual to physical address mapping is accomplished through a random first-touch translation mechanism, enabling long-running simulations [7]. We estimate the delay of the caches using CACTI 7.0 [8]. The cache block size is 64 bytes in the entire memory hierarchy.

Table 2 summarizes the key characteristics of our simulated workloads which consist of server and SPEC [10] programs. We use the SimFlex [11] methodology to simulate server workloads. For every server application, we create five checkpoints with warmed caches, branch predictors, and prediction tables. Each checkpoint is drawn over an interval of 10 s of simulated time as observed by the OS. Then we run 200 K instructions from each checkpoint, using the first 40 K instructions for warming queues (e.g., ROB), and the rest for actual measurements. Based on SimFlex [11], our measurements are computed with 95% confidence and less than 4% error. For SPEC benchmarks, we run the simulations for at least 100 M instructions on every core and use the first 20 M as the warm-up and the next 80 M for measurements.

Table 2 Spatial prefetching: Application parameters.

Application	Description	LLC MPKI
Server		
Data Serving	Cassandra Database, 15GB Yahoo! Benchmark	6.7
SAT Solver	Cloud9 Parallel Symbolic Execution Engine	1.7
Streaming	Darwin Streaming Server, 7500 Clients	3.9
Zeus	Zeus Web Server v4.3, 16 K Connections	5.2
Scientific		
em3d	400 K Nodes, Degree 2, Span 5, 15% Remote	32.4
SPEC		
Mix 1	lbm, omnetpp, soplex, sphinx3	15.7
Mix 2	lbm, libquantum, sphinx3, zeusmp	12.5
Mix 3	milc, omnetpp, perlbench, soplex	12.7
Mix 4	astar, omnetpp, soplex, tonto	14.7
Mix 5	GemsFDTD, gromacs, omnetpp, soplex	12.6

We set the spatial region size to 2 KB. We configure SMS and Bingo with 16 K-entry 16-way associative PHTs. We configure VLDP with a 16-entry Delta History Buffer, 64-entry Offset Prediction Table, and three 64-entry Delta Prediction Tables. We study all data prefetchers in the context of the LLC. We consider every core to have its own prefetcher, independent of others. All methods are triggered upon LLC accesses and prefetch directly into the LLC (i.e., no prefetch buffer).

2.2 Results

To evaluate the effectiveness of the proposed prefetcher, Fig. 1 shows the coverage and overprediction of the competing prefetching techniques. As shown, Bingo offers the highest miss coverage across all workloads. By associating footprint metadata to more than one event, and matching the longest event, Bingo maximally and precisely extracts spatially-correlated data access patterns and significantly reduces the number of cache misses. On average, Bingo covers more than 63% of cache misses, outperforming the second-best prefetcher by 8%. The overprediction of Bingo is on par with the rest of the competing prefetchers.

Fig. 2 shows the performance improvement of Bingo along with other prefetching techniques, over a baseline without a prefetcher. Bingo consistently outperforms the competing prefetching approaches across all workloads. The performance improvement of Bingo ranges from 11% in Zeus to 285% in em3d. Miss coverage, timeliness, and the accuracy of prefetches are the main contributors to Bingo's superior performance improvement. For most of the workloads, Bingo provides a significant performance improvement. In Zeus, however, memory accesses are more

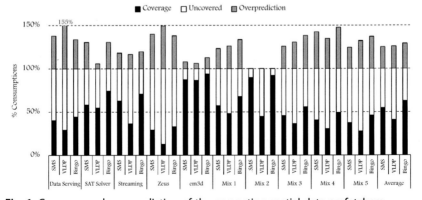

Fig. 1 Coverage and overprediction of the competing spatial data prefetchers.

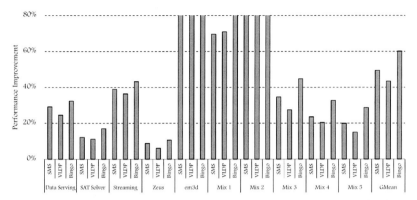

Fig. 2 Performance comparison of spatial prefetching techniques, normalized to a baseline system with no prefetcher.

temporally correlated than spatially [3]. Even those accesses that are spatially predictable are already fetched in parallel by the out-of-order processing, resulting in a negligible performance improvement with spatial prefetchers.

3. Temporal prefetching

3.1 Methodology

We evaluate temporal prefetchers, STMS [12], ISB [13], and Domino [14]. Table 3 summarizes key elements of our methodology.

The chip is modeled based on a quad-core processor with 4 MB of last-level cache (LLC). The cache hierarchy of each core includes a

Table 3 Temporal prefetching: Evaluation parameters.

Parameter	Value
Chip	Four cores, 4 GHz
Core	SPARC v9 ISA, 8-stage pipeline, out-of-order execution, 4-wide issue, 128-entry ROB, 64-entry LSQ
I-Fetch Unit	64 KB, 2-way, 2-cycle load-to-use, next-line prefetcher, hybrid branch predictor, 16 K gShare & 16 K bimodal
L1-D Cache	64 KB, 2-way, 2-cycle load-to-use, 4 ports, 32 MSHRs
L2 Cache	4 MB, 16-way, 18-cycle hit latency, 64 MSHRs
Memory	45 ns delay, 37.5 GB/s peak bandwidth

64 KB data and a 64 KB instruction cache. The 4 MB LLC is distributed among four slices. Cache line size is 64 bytes. The chip has two memory controllers that provide up to 37.5 GB/s of off-chip bandwidth.

We simulate systems running *Solaris* and executing the workloads listed in Table 4. We include a variety of server workloads from competing vendors, including online transaction processing, CloudSuite [15], and Web server benchmarks. Prior work [16] has shown that these workloads have characteristics representative of the broad class of server workloads.

We use a combination of trace-based and timing full-system simulations to evaluate our proposal. Our trace-based analyses use traces collected from Flexus [17] with in-order execution, no memory system stalls, and a fixed instruction-per-cycle (IPC) of 1.0.

We estimate the performance of various designs using Flexus full-system timing simulation. Flexus timing simulator extends the Virtutech Simics functional simulator with timing models of cores, caches, on-chip protocol controllers, and interconnect. Flexus models the SPARC v9 ISA and is able to run unmodified operating systems and applications.

Table 4 Temporal prefetching: Application parameters.
CloudSuite

Data Serving	Cassandra 0.7.3 Database 15GB Yahoo! Cloud Serving Benchmark
MapReduce-C	Hadoop 0.20.2 Bayesian Classification Algorithm
MapReduce-W	Hadoop 0.20.2 Mahout 0.4 Library
SAT Solver	Cloud9 Parallel Symbolic Execution Engine Four 5-byte and One 10-byte Arguments
Media Streaming	Darwin Streaming Server 6.0.37500 Clients, 60 GB Dataset, High Bitrate
Web Search	Nutch 1.2/Lucene 3.0.1, 230 Clients 1.4 GB Index, 15 GB Data Segment
Web Server (SPECweb99)	
Web Apache	Apache HTTP Server v2.0, 16 K Connections FastCGI, Worker Threading Model
Web Zeus	Zeus Web Server v4.3 16 K Connections, FastCGI
Online Transaction Processing (TPC-C)	
OLTP	Oracle 10 g Enterprise Database Server 100 Warehouses (10 GB), 1.4 GB SGA

We use the SimFlex multiprocessor sampling methodology [11]. For each measurement, we launch simulations from checkpoints with warmed caches and branch predictors and run 300 K cycles to achieve a steady state of detailed cycle-accurate simulation before collecting measurements for the subsequent 150 K cycles. We use the ratio of the number of application instructions to the total number of cycles (including the cycles spent executing operating system code) to measure performance; this metric has been shown to accurately reflect overall system throughput of multiprocessors [11]. Performance measurements are computed with 95% confidence and an error of less than 4%.

We consider infinite-size metadata tables for STMS and ISB. We set the EIT and History Table of Domino to 2 and 16 M entries, respectively. To have a fair evaluation, all prefetchers are trained using L1-D miss sequences, and all prefetchers prefetch into a small prefetch buffer near the L1-D cache, which contains 32 cache blocks. The degree of prefetching for all prefetchers is set to four.

3.2 Results

Fig. 3 shows the coverage and overprediction for the competing prefetching techniques. Domino outperforms other prefetching techniques in term of covering more cache misses. The second best prefetcher is STMS. In all workloads, Domino either significantly outperforms STMS (19% in *OLTP*) or at isocoverage, offers substantially higher accuracy. On average, Domino's discards come within nearly one-third of that of STMS. Corroborating prior work [18], our results show that PC-localized temporal prefetchers like ISB are not useful in the context of server workloads. PC

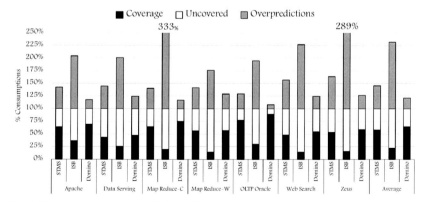

Fig. 3 Coverage and overprediction of the competing temporal data prefetchers.

localization temporal prefetching suffers from two fundamental obstacles: (1) PC localization tears the strong temporal correlation between global miss addresses, and (2) PC localization prefetches the following misses of a PC, which may not be the subsequent misses of the program. Since server workloads have extensive instruction working sets [16], the re-execution of a specific load instruction in the execution sequence may take a long time. Therefore, the prefetched blocks may be evicted before re-execution of the load instruction.

Fig. 4 shows the performance improvement of Domino prefetcher along with ISB and STMS, over a baseline with no prefetcher. The figure clearly shows the ability of Domino prefetcher in boosting performance.

In 8 out of 9 workloads, Domino outperforms other temporal prefetchers due to its higher coverage and/or better timeliness. The average performance improvement of Domino prefetcher over the baseline is 16%. The second-best prefetcher is STMS with the average performance improvement of 10%.

For most of the workloads, Domino provides a significant performance improvement. *Web Search* and *Media Streaming* have relatively high MLP, and hence, many of the misses that prefetchers capture, are already fetched in parallel with out-of-order execution. Therefore, despite high coverage, prefetchers are unable to boost the performance of these workloads significantly. In *MapReduce-W,* temporal streams identified by methods are drastically short, and hence, the delay of fetching metadata from memory does not get amortized among subsequent prefetches, resulting in less performance enhancement. *SAT Solver* produces its dataset on-the-fly during

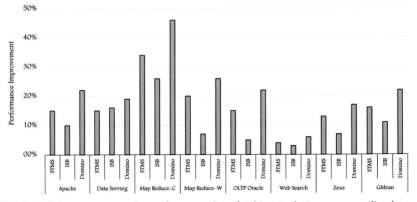

Fig. 4 Performance comparison of temporal prefetching techniques, normalized to a baseline system with no prefetcher.

execution, and thus, does not have a static and well-structured dataset [19]. Consequently, its memory accesses are hard-to-predict and all methods manifest low coverage and high overprediction, and accordingly, small performance improvement.

4. Spatio-temporal prefetching

4.1 Methodology

We evaluate STeMS [20], a spatio-temporal prefetcher, against state-of-the-art spatial and temporal prefetchers including SMS [1], VLDP [2], STMS [12], and ISB [13]. Table 5 summarizes the key elements of our methodology.

Our platform, which is modeled after Intel Xeon Processor™ [21], is a quad-core processor with 4 MB of last-level cache and two memory channels. Core microarchitecture includes 32 KB L1-D and L1-I caches. The cache block size is 64 bytes in the entire memory hierarchy. Two memory channels are located at the corner of the chip and provide up to 37.5 GB/s of off-chip bandwidth. We use CACTI [22] to estimate the delay of on-chip caches.

We simulate systems running Solaris and executing the workloads listed in Table 6. We include a variety of server workloads from competing vendors, including Online Transaction Processing, Decision Support System, Web Server, and CloudSuite [15].

We use trace-driven simulations for evaluating the ability of each prefetcher at predicting future cache misses and timing simulations for

Table 5 Spatial, temporal, and spatio-temporal prefetching: Evaluation parameters.

Parameter	Value
Chip	14 nm, 4 GHz, 4 cores
Core	Out-of-order execution, 4-wide dispatch/retirement, 64-entry LSQ, 256-entry ROB
I-Fetch Unit	32 KB, 2-way, 2-cycle load-to-use, 24-entry pre-dispatch queue, Perceptron branch predictor [9]
L1-D Cache	32 KB, 2-way, 2-cycle load-to-use
L2 Cache	1 MB per core, 16-way, 11 cycle lookup delay, 128 MSHRs
Memory	240-cycle delay, 37.5 GB/s peak bandwidth, two memory controllers

Table 6 Spatial, temporal, and spatio-temporal prefetching: Application parameters.

OLTP—Online Transaction Processing (TPC-C)

DB2	**IBM DB2 v8 ESE, 100 warehouses (10 GB), 2 GB buffer pool**

DSS—Decision Support Systems (TPC-H)

Qry 2 and 17	IBM DB2 v8 ESE, 480 MB buffer pool, 1 GB database

Web Server (SPECweb99)

Apache	Apache HTTP server v2.0, 16 K connections, fastCGI, worker threading
Zeus	Zeus web server v4.3, 16 K connections, fastCGI

CloudSuite

Data Serving	Cassandra 0.7.3 Database, 15 GB Yahoo! Cloud Serving Benchmark
MapReduce	Hadoop 0.20.2, Bayesian classification algorithm
Media Streaming	Darwin Streaming Server 6.0.3, 7500 Clients, 60 GB dataset, high bitrates
Web Search	Nutch 1.2/Lucene 3.0.1, 230 Clients, 1.4 GB Index, 15 GB Data Segment

performance studies. The trace-driven experiments use traces that are obtained from the in-order execution of applications with a fixed instruction-per-cycle (IPC) of 1.0. We run trace-driven simulations for 12 billion instructions and use the first half as the warmup and the rest for the actual measurements. We use ChampSim full-system simulator [4] for timing experiments. For every application, we create five checkpoints with warmed architectural components (e.g., branch predictors, caches, and prediction tables). Each checkpoint is drawn over an interval of 10 s of simulated time as observed by the Operating System (OS). We execute 200 K instructions from each checkpoint and use the first 40 K instructions for warm-up and the rest for measurements.

For every prefetcher, we do a sensitivity analysis to find the storage requirement for the prefetcher to have a reasonable miss coverage. For most of the prefetchers, we begin with an infinite storage and reduce the area until the miss coverage drops more than 5% as compared to its peak value; doing so, we ensure that every prefetcher is able to provide its maximum possible performance improvement, negating the limiting effect of dissimilar storage requirements of different prefetchers, enabling fair comparison among various approaches.

To have a fair comparison, we consider the following items in the implementation of the competing approaches:

• All prefetchers are trained on L1-D misses (LLC accesses). This way, each core has its own prefetcher and issues prefetch requests for itself, independent of others.

• Except for STMS and STeMS, other prefetching techniques directly prefetch into the primary data cache (L1-D). STMS and STeMS rely on temporal correlation of global misses and could be inaccurate in some cases. Consequently, streaming the prefetched data into the L1-D may significantly pollute the cache. For STMS and STeMS, we place the prefetched data in a small buffer next to L1-D cache. It is worth mentioning that the other prefetching techniques do *not* benefit from such an approach (i.e., prefetching into a small prefetch buffer). This is because of the fact that the other prefetching techniques do not prefetch the *next miss in time* and usually prefetch cache blocks that a program may need far from the current miss. Therefore, prefetching into a small prefetch buffer would result in the early eviction of the prefetched blocks, diminishing the performance.

• The prefetching lookahead of all methods, except SMS, is set to four, providing a sensible trade-off between the performance improvement and the off-chip bandwidth overhead. The prefetching lookahead of SMS depends on the recorded patterns and varies across regions. As SMS does *not* keep the order of prefetch candidates, it is not possible to enforce a predefined prefetching lookahead to it.

We simulate the competing data prefetchers with the following configurations. As a point of reference, we also include the *opportunity* results for temporal, spatial, and spatio-temporal data prefetching techniques. **Sampled Temporal Memory Streaming (STMS):** STMS uses a 6-million-entry Index Table and a 6-million-entry History Table. The Index Table is indexed by the lower bits of the trigger address.

Irregular Stream Buffer (ISB): Two 4 K-entry on-chip structures are used to cache the content of two 3 M-entry off-chip metadata tables. Parts of the metadata that may be used by the processor are always in the on-chip structures.

Spatial Memory Streaming (SMS): Our sensitivity analysis demonstrates that 16 K-entry PHT is sufficient for SMS to reach the peak miss coverage. The memory space is divided into 2 KB Spatial Regions.

Variable Length Delta Prefetcher (VLDP): VLDP is equipped with a 16-entry DHB, 64-entry OPT, and three 128-entry fully-associative DPTs. The size of spatial regions is set to 2 KB.

Spatio-Temporal Memory Streaming (STeMS): STeMS uses a 2 M-entry RMOB, a 2 M-entry Index Table, and a 16 K-entry PST. The indexing scheme of metadata tables is exactly the same as that of STMS and SMS. STeMS uses the on-chip structure of STMS, with the same configuration, for reading and updating the RMOB and the Index Table entries.

We run trace-based simulations for profiling and miss coverage studies and detailed cycle-accurate timing simulations for performance experiments.

4.2 Results

To compare the effectiveness of prefetching techniques, Fig. 5 shows the coverage and overprediction of the competing prefetching techniques. Our results show that STMS outperforms ISB regarding both miss coverage and the overprediction. We find that *PC localization* is the main source of the inefficiency of ISB. ISB correlates temporal streams with load instructions and aggressively issues prefetch requests for the future references of the load instructions. We find that this mechanism suffers from two main obstacles: (1) PC-localization breaks the strong temporal correlation among global addresses that is dominant in server workloads, and (2) PC-localization prefetches the next misses of a load instruction, which may not be the next misses of the program. As server workloads have large instruction working sets (in the range of few megabytes [16]), the re-execution of a specific load instruction in the execution order may take a long time; hence, the prefetched cache blocks of a PC-localized temporal

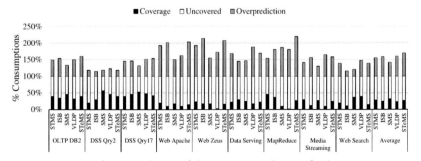

Fig. 5 Coverage and overprediction of the competing data prefetchers.

prefetcher may be evicted from the cache due to conflicts, prior to the re-execution of the load instruction. While STMS captures more cache misses than ISB, it falls short of covering the entire temporal opportunity.

SMS covers more cache misses than VLDP with lower overprediction rate. Our investigations show that the *multi-degree* prefetching mechanism (i.e., increasing the prefetching lookahead beyond one) of VLDP is the most contributor to its large overprediction rate. Once VLDP has predicted the next access of a spatial region, it makes more prefetches using the prediction as an input to the metadata tables. We find that this strategy is inaccurate for server workloads, and gets worse as VLDP further repeats this process. Despite high overhead (i.e., large tables and bandwidth overhead), STeMS covers only 28% of the opportunity, making it quite ineffective.

Figure 6 shows the performance improvement of the competing prefetchers over a baseline with no data prefetcher. STMS outperforms ISB thanks to its higher coverage and accuracy. However, STMS suffers from a high start-up cost, as the first prefetch request is sent after two memory round-trip latency. This is why notwithstanding offering high miss coverage, STMS is unable to significantly boost the performance of applications that are dominated with short temporal streams (e.g., Web Search).

SMS outperforms VLDP due to its higher coverage and lower overprediction rate. While SMS and VLDP are effective at boosting the performance in most of the evaluated workloads, they offer little benefits for the MapReduce workload. We find that most of the misses that SMS and VLDP cover in this workload are independent misses, and hence, they are already fetched in parallel with the out-of-order execution.

The performance improvement of STeMS ranges from 1% in DSS Qry2 to 18% in Data Serving. Like STMS, the metadata of STeMS is located off-the-chip, and hence, STeMS suffers from a high start-up latency. STeMS can start issuing prefetch requests only after three serial memory

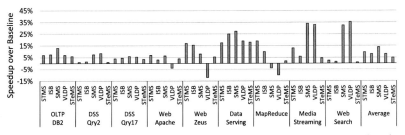

Fig. 6 Performance comparison of prefetching techniques, normalized to a baseline system with no prefetcher.

round-trip access latencies. This makes STeMS ineffective, especially for applications that are dominated by *sparse* spatial regions (i.e., regions that have few predicted blocks). In such regions, the start-up latency of prefetching is not compensated because only few cache blocks are prefetched after waiting a long time for the metadata to arrive.

5. Offset prefetching

5.1 Methodology

We evaluate MLOP [23] and BOP [24] in the context of the simulation framework provided with Third Data Prefetching Championship (DPC-3) [25]. We follow the evaluation methodology of the championship and run simulations for all 46 provided single-core traces from SPEC workloads [10]. Out of 46 provided traces, we exclude two non-memory-insensitive ones, then create 11 random MIX traces from the other 44 (the MIXes are completely different from each other; no single-core trace repeats in two of the MIXes). For better readability, we only report the simulation results for workloads whose performance is highly affected by the evaluated prefetchers, as well as the average of all simulated workloads. Both prefetchers sit in the L1 data cache and are trained by L1-D miss streams.

Figs. 7 and 8 show the miss coverage and performance results, respectively. We report miss coverage results only for single-core programs and performance results for all simulated workloads.

MLOP offers the higher miss coverage and performance improvement than BOP. On average, MLOP covers 56% of cache misses, X% more than BOP.

The performance analysis shows that MLOP outperforms BOP on both single-core and multi-core platforms. On average, MLOP improves performance by 30%, outperforming BOP by X%.

Miss coverage and timeliness are the main contributors to MLOP's superior performance improvement. BOP, due to its binary classification, neglects many appropriate prefetch offsets and hence, falls short of covering a significant fraction of cache misses. Moreover, we find that another deficiency of BOP arises from the fact that it updates merely a single offset in each update; whereas, MLOP uses vector operations to efficiently update all offset scores at once. MLOP, by considering both miss coverage and timeliness at evaluation and selection of its prefetch offsets, provides the best of both worlds, significantly improving miss coverage and timeliness of prefetching, thereby providing significant performance benefits.

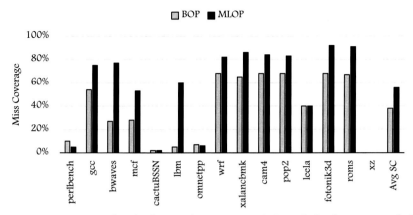

Fig. 7 Miss coverage of prefetching techniques. "Avg SC" stands for the average of all single-core workloads.

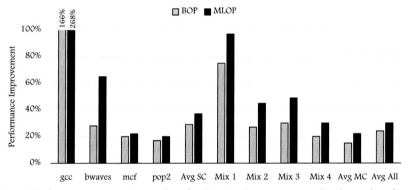

Fig. 8 Performance comparison of prefetching techniques, normalized to a baseline system with no prefetcher. Mix1 = {mcf_s-472B, gcc_s-1850B, roms_s-1070B, cam4_s-490B}, Mix2 = {xz_s-2302B, mcf_s-484B, fotonik3d_s-8225B, bwaves_s-1740B}, Mix3 = {pop2_s-17B, roms_s-1613B, bwaves_s-2609B lbm_s-4268B}, and Mix4 = {roms_s-1007B, fotonik3d_s-7084B, mcf_s-1554B, xalancbmk_s-165B}. "Avg SC/MC/All" stands for the average of single-core/multi-core/all workloads.

6. Multi-degree prefetching with pairwise-correlating prefetchers

6.1 Methodology

We evaluate RMD [26] and asses its effectiveness over VLDP [2], a pairwise-correlating data prefetcher. We use ChampSim [4] to simulate a system whose configuration is shown in Table 7. We use SPEC 2006 benchmark [10] for our evaluations. For single-core experiments, we report the results for 15 memory-intensive programs; we create MIX workloads

Table 7 Pairwise-correlating prefetching: Evaluation parameters.

Parameter	Value
Processor	4 cores, 8-wide OoO
L1-I/D	32 KB, 8-way set-associative, 1-cycle load-to-use, 64 B lines
L2 Cache	1 MB per core, 16-way set-associative, unified, 11-cycle access latency
Memory	DDR4-2133MHz, 2 ranks/channel, 8 banks/rank, 2 KB row buffer/ bank, tCL-tRCD-tRP-tRAS = 15-15-15-39

of these programs for our multi-core evaluations. The memory channel is modeled based on borrowed data from commercial DDR4-2133 technology specification [27], which provides peak bandwidth of around 17,064 MB/s. We run the simulations for at least 100 M instructions on every core and use the first 20 M as the warm-up and the rest for actual measurements. All data prefetchers are implemented near L1 data cache.

6.2 Results

Fig. 9 shows miss coverage and overpredictions of competing methods. Moreover, numbers on the bars indicate the performance of methods, normalized to a baseline system with no prefetcher.

As Fig. 9 shows, *RMD* significantly reduces the overpredictions of *VLDP*. The reduction for *VLDP* is 1.4× on average and up to 3.6×. The large reduction in overpredictions results in substantially less memory bandwidth consumption and cache pollution, which are crucially important for performance and energy-efficiency especially in multi- and many-core

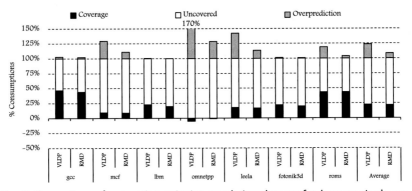

Fig. 9 Comparison of competing pairwise-correlating data prefetchers on single-core workloads.

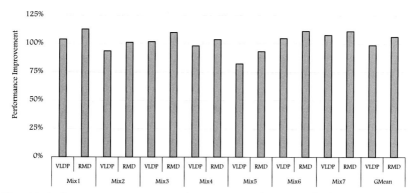

Fig. 10 Performance comparison of pairwise-correlating prefetching techniques, normalized to a baseline system with no prefetcher on multicore workloads.

substrates. While offering large reductions in mispredictions, *RMD* reduces the miss coverage negligibly; with *RMD*, *VLDP* offer only 1.9% lower miss coverage as compared to without *RMD*. The performance with *RMD* is on par with that of without it. Note that, single-core substrates typically are *not* bandwidth-limited [28], and that is why the reduction in the overprediction does not translate into performance enhancement.

Fig. 10 shows the performance of all methods normalized to a baseline with no data prefetcher. The evaluated workloads in this figure are four-core MIX workloads, where MIX$_i$ is created by picking programs (i, $i+1$, $i+2$, $i+3$)%15 (modulo 15) from Fig. 9. As the results show, augmenting *VLDP* by *RMD* results in significant performance improvements. The performance improvement of *RMD* on top of *VLDP* is 5.8% on average and up to 15%.

7. Summary

Data prefetching is a historical approach to reduce the number of cache misses, thereby increasing system performance. Different data prefetchers could have unalike effectiveness on different applications. In this chapter, we evaluate state-of-the-art prefetchers on a wide variety of applications and show how, why, and to what extent they are effective in improving performance.

References
[1] S. Somogyi, T.F. Wenisch, A. Ailamaki, B. Falsafi, A. Moshovos, Spatial memory streaming, in: *Proceedings of the International Symposium on Computer Architecture (ISCA)*, 2006, pp. 252–263.

[2] M. Shevgoor, S. Koladiya, R. Balasubramonian, C. Wilkerson, S.H. Pugsley, Z. Chishti, Efficiently prefetching complex address patterns, in: *Proceedings of the International Symposium on Microarchitecture (MICRO)*, 2015, pp. 141–152.

[3] M. Bakhshalipour, M. Shakerinava, P. Lotfi-Kamran, H. Sarbazi-Azad, Bingo spatial data prefetcher, in: *Proceedings of the International Symposium on High-Performance Computer Architecture (HPCA)*, 2019, pp. 399–411.

[4] ChampSim, 2017. https://github.com/ChampSim/.

[5] S. Pugsley, A. Alameldeen, C. Wilkerson, H. Kim, The Second Data Prefetching Championship (DPC-2), IEEE, 2015.

[6] Intel Xeon Processor E3-1220 v6, 2017. https://www.intel.com/content/www/us/en/products/processors/xeon/e3-processors/e3-1220-v6.html/.

[7] S. Franey, M. Lipasti, Tag tables, in: *Proceedings of the International Symposium on High-Performance Computer Architecture (HPCA)*, 2015, pp. 514–525.

[8] CACTI 7.0: A Tool to Model Caches/Memories, 3D Stacking, and Off-Chip IO, 2017. https://github.com/HewlettPackard/cacti/.

[9] D.A. Jiménez, C. Lin, Dynamic branch prediction with perceptrons, in: *Proceedings of the International Symposium on High-Performance Computer Architecture (HPCA)*, 2001, pp. 197–206.

[10] J.L. Henning, SPEC CPU2006 benchmark descriptions, ACM SIGARCH Comput. Archit. News 34 (4) (2006) 1–17.

[11] T.F. Wenisch, R.E. Wunderlich, M. Ferdman, A. Ailamaki, B. Falsafi, J.C. Hoe, SimFlex: statistical sampling of computer system simulation, IEEE Micro 26 (2006) 18–31.

[12] T.F. Wenisch, M. Ferdman, A. Ailamaki, B. Falsafi, A. Moshovos, Practical off-chip meta-data for temporal memory streaming, in: Proceedings of the International Symposium on High Performance Computer Architecture (HPCA), 2009, pp. 79–90.

[13] A. Jain, C. Lin, Linearizing irregular memory accesses for improved correlated prefetching, in: Proceedings of the International Symposium on Microarchitecture (MICRO), 2013, pp. 247–259.

[14] M. Bakhshalipour, P. Lotfi-Kamran, H. Sarbazi-Azad, Domino temporal data prefetcher, in: Proceedings of the International Symposium on High-Performance Computer Architecture (HPCA), IEEE, 2018, pp. 131–142.

[15] CloudSuite, 2012. Available at http://cloudsuite.ch.

[16] M. Ferdman, A. Adileh, O. Kocberber, S. Volos, M. Alisafaee, D. Jevdjic, C. Kaynak, A.D. Popescu, A. Ailamaki, B. Falsafi, Clearing the clouds: a study of emerging scale-out workloads on modern hardware, in: Proceedings of the International Conference on Architectural Support for Programming Languages and Operating Systems (ASPLOS), 2012, pp. 37–48.

[17] Flexus, 2012. http://parsa.epfl.ch/simflex.

[18] T.F. Wenisch, Temporal Memory Streaming, PhD thesis, Carnegie Mellon University, 2007.

[19] D. Jevdjic, S. Volos, B. Falsafi, Die-stacked DRAM caches for servers: hit ratio, latency, or bandwidth? Have it all with footprint cache, in: Proceedings of the International Symposium on Computer Architecture (ISCA), 2013, pp. 404–415.

[20] S. Somogyi, T.F. Wenisch, A. Ailamaki, B. Falsafi, Spatio-temporal memory streaming, in: Proceedings of the International Symposium on Computer Architecture (ISCA), 2009, pp. 69–80.

[21] Intel Xeon Processor E3-1245 v6, 2017. Available at https://www.intel.com/content/www/us/en/products/processors/xeon/e3-processors/e3-1245-v6.html.

[22] N. Muralimanohar, R. Balasubramonian, N. Jouppi, Optimizing NUCA organizations and wiring alternatives for large caches with CACTI 6.0, in: *Proceedings of the International Symposium on Microarchitecture (MICRO)*, 2007, pp. 3–14.

[23] M. Shakerinava, M. Bakhshalipour, P. Lotfi-Kamran, H. Sarbazi-Azad, Multi-lookahead offset prefetching, in: The Third Data Prefetching Championship, IEEE, 2019.

[24] P. Michaud, Best-offset hardware prefetching, in: *Proceedings of the International Symposium on High-Performance Computer Architecture (HPCA)*, 2016, pp. 469–480.

[25] The Third Data Prefetching Championship, 2019. https://dpc3.compas.cs.stonybrook.edu/.

[26] F. Golshan, M. Bakhshalipour, M. Shakerinava, A. Ansari, P. Lotfi-Kamran, H. Sarbazi-Azad, Harnessing pairwise-correlating data prefetching with runahead metadata, IEEE Comput. Archit. Lett. 19 (2) (2020) 130–133, https://doi.org/10.1109/LCA.2020.3019343.

[27] JEDEC-DDR4. n.d. https://www.jedec.org/sites/default/files/docs/JESD79-4.pdf.

[28] B.M. Rogers, A. Krishna, G.B. Bell, K. Vu, X. Jiang, Y. Solihin, Scaling the bandwidth wall: challenges in and avenues for CMP scaling, in: *Proceedings of the International Symposium on Computer Architecture (ISCA)*, 2009, pp. 371–382.

About the authors

Mehran Shakerinava received his B.Sc. in Computer Engineering from Sharif University of Technology, Iran (2019). He is currently pursuing a Ph.D. in Computer Science at McGill University and Mila Quebec AI Institute.

Fatemeh Golshan received the B.Sc. degree in Computer Engineering from Isfahan University of Technology, Isfahan, Iran, in 2018, and the M.Sc. degree in Computer Architecture from the Sharif University of Technology, Tehran, Iran, in 2020. Her research interest is Computer Architecture with an emphasis on memory systems.

Ali Ansari is a Ph.D student at the École Polytechnique Fédérale de Lausanne (EPFL) pursuing his program under the supervision of professor Babak Falsafi. He graduated from the master's program from the Sharif University of Technology in Tehran, Iran, in 2019. During his master studies, he researched instruction and data prefetching techniques advised by professors Hamid Sarbazi-Azad and Pejman Lotfi-Kamran. His research interests span high-performance computer architecture, memory systems, and emerging workloads.

Pejman Lotfi-Kamran is an associate professor and the head of the School of Computer Science and the director of Turin Cloud Services at Institute for Research in Fundamental Sciences (IPM). His research interests include computer architecture, computer systems, approximate computing, and cloud computing. His work on scale-out server processor design lays the foundation for Cavium ThunderX. Lotfi-Kamran has a Ph.D. in Computer Science from the École Polytechnique Fédérale de Lausanne (EPFL). He received his MS and BS in Computer Engineering from the University of Tehran. He is a member of the IEEE and the ACM.

Hamid Sarbazi-Azad is currently a professor of Computer Science and Engineering at the Sharif University of Technology, Tehran, Iran. His research interests include high-performance computer/memory architectures, NoCs and SoCs, parallel and distributed systems, social networks, and storage systems, on which he has published over 400 refereed papers. He received Khwarizmi International Award in 2006, TWAS Young Scientist Award in Engineering Sciences in 2007, Sharif University Distinguished Researcher awards in the years 2004, 2007, 2008, 2010, and 2013, the Iranian Ministry of Communication and Information Technology's award for contribution to IT research and education in 2014, and distinguished book author of the Sharif University of Technology in 2015 and 2020. Dr Sarbazi-Azad is now an associate editor of ACM Computing Surveys, IEEE Computer Architecture Letters, and Elsevier's Computers and Electrical Engineering.

Printed in the United States
by Baker & Taylor Publisher Services